J. E. (John Ellor) Taylor

Mountain and Moor

J. E. (John Ellor) Taylor

Mountain and Moor

ISBN/EAN: 9783743332454

Manufactured in Europe, USA, Canada, Australia, Japa

Cover: Foto ©berggeist007 / pixelio.de

Manufactured and distributed by brebook publishing software
(www.brebook.com)

J. E. (John Ellor) Taylor

Mountain and Moor

NATURAL HISTORY RAMBLES.

MOUNTAIN AND MOOR.

BY

J. E. TAYLOR, F.L.S., F.G.S.,

EDITOR OF "SCIENCE-GOSSIP," ETC.

PUBLISHED UNDER THE DIRECTION OF
THE COMMITTEE OF GENERAL LITERATURE AND EDUCATION
APPOINTED BY THE SOCIETY FOR PROMOTING
CHRISTIAN KNOWLEDGE.

SOCIETY FOR PROMOTING CHRISTIAN KNOWLEDGE:
LONDON: 77, GREAT QUEEN ST., LINCOLN'S-INN FIELDS;
4, ROYAL EXCHANGE; 48, PICCADILLY;
AND BY ALL BOOKSELLERS.
NEW YORK: POTT, YOUNG, & CO.

1879.

WYMAN AND SONS, PRINTERS,
GREAT QUEEN STREET, LINCOLN'S INN FIELDS,
LONDON, W.C.

CONTENTS.

MOUNTAIN AND MOOR.

CHAPTER I.

OUR BRITISH MOUNTAINS—THEIR GEOLOGICAL ANTIQUITY.

The High Mountain Ranges of the World—The Andes, Himalayas, Alps, &c.—Our British Mountains much older Geologically—Mountains and Literature—Comparison of the Ages of Sutherland Mountains and Himalayas—Evidences of Metamorphism, &c., in Strata of our Mountains—Evidences of Denudation from British Mountains—Skiddaw, and the Western Highlands—Suilven—Mountains of North Wales—Pentland Hills, the Grampians, Mendips, Quantock Hills, &c.—Silurian Hills and Mountains—Carboniferous Hills—Peak of Derbyshire, Ingleborough Fell—Millstone Grit Hills—Penyghent, Kinder Scout—The Pennine Chain—Hills of New Red Sandstone—Beeston Hill—Cotteswold and Cleveland Hills—Chalk Wolds and Downs—Age of the Wrekin—Ancient Volcanoes of Great Britain—The Isle of Mull—Rev. J. Clifton-Ward on Ancient Cumberland Volcanoes, and Denudation of Lake District.

THERE are other mountains and hills in the world than those occurring in Great Britain. Some of them stand base to base, and extend over many degrees of latitude or longitude, and we term them "mountain-chains." Their highest peaks penetrate far into the clouds, and it may be that the snow-line enshrouds

B

nearly half their heights, as it does the Himalayas.
Human beings could hardly bear the decreased atmospheric pressure which is characteristic of Mount
Everest, the highest point of the Himalayan mountains, five and a half miles above the sea-level. No
wonder that the loftily-elevated masses of the earth's
rocky crust forming the Himalayas should be snow-
clad, and so have originated the native name, which
signifies the " Abode of Snow."

Our British mountains are not marked by the presence of active volcanoes, like those which crown the
peaks of the Andes, and which have built the latter
up still higher by the heaps of rubbish and lava they
have accumulated around their craters. They are
not mountains which tremble and groan, like Sinai of
old, beneath the pressure of volcanic forces ; or whose
flanks are ascended with suspicion of a possible
overwhelming beneath clouds of hot ashes, as with
Vesuvius and Etna. They are neither affected by
fire nor ice. No volcanic flames light up their
summits, no glaciers now glide down their rugged
sides. The numerous ice-rivers which give to the
Swiss Alps half their natural charms are only represented on the loftiest and coldest of our British
mountains, such as Ben Nevis, even in its darkest and
most protected fissures, by the snow-wreath which
has lingered since last winter !

And yet our British mountains have a history tran-
scending in geological and general interest those of
any other part of the globe, no matter how the latter
overtower them in loftiness or in the importance of
their prevailing physical phenomena. Many of them

have been the sport alternately, and almost pendulum-like, of the volcano and the glacier. They have been crushed under continental ice, like the unknown and as yet unseen mountains of Greenland; whilst some of the most notable of them owe the very existence of the masses which remain after millions of years' wear-and-tear, to the materials ejected from ancient volcanoes on or near their present sites! Such hills and mountains are the Wrekin, Mount Sorrel, Snowdon, and many of the Cumberland and Scotch mountains.

No other mountains in the world, except the hills of Greece, have such rich historical and poetical associations as those of our own country. Most mountain regions have sheltered the defenders of human freedom in all ages from the overwhelming arm of the oppressor; and British hilly fortresses have sometimes served in as good stead as similar mountain homes to other lovers of freedom! The hills and mountain slopes, where one seems to breathe faster and freer, and to be more removed from the artificial restraints of society, have consoled many a fugitive from despotic oppression, and have been the home of brave lovers of liberty, long after the plains have yielded to the yoke of the tyrant. The literature of almost every mountain country in the world is a patriotic literature. Liberty always found her choicest home in the mountains—long even before the time when David fled from the wrath of Saul. Our own English mountains fortunately have had little opportunity thus to influence our history. But those of Ireland, Wales, and of the Scottish High-

lands have given birth to many a stirring lyric and
ballad. The genius of Scott has sanctified the bleak
hills of Scotland, and that of Wordsworth has thrown
an additional halo of peaceful beauty about the
mountains of the Lake district of England.

But it is of the rock masses, upon which the hand
of Time has been employed for ages in sculpturing
them into their present hilly and mountainous forms
and shapes, that we prefer to speak. We have seen
that as regards height and extent the mountain chains
of the Urals, the Alps, Pyrenees, Himalayas, Andes,
&c., far exceed them. But nearly all these mountains are but as of yesterday in comparison with the
age of some of our chief British hills and mountains.
Had they been exposed for the same length of time
to the wear-and-tear of atmospheric denudation, it is
doubtful whether they would have been now of as
great height as ours are. For these British hills were
much higher than they are at present long before the
materials of which the Alps are composed had been
gathered together. The seas, along whose floors the
marine sediments were accumulated out of which the
rocks of the Himalayas are formed, had not even
come into existence when the existing high mountains
of Sutherland raised their peaks above the clouds to
an altitude greatly exceeding that of our own day.

In short, there is no fact in modern geology better
known than that different mountains are formed of
rocks which were accumulated originally as sediments, at different periods of our world's past history. Not only were the rock materials of the mountains formed at different times, they were upheaved

to even more than their present heights at different periods also. Thus, from a knowledge of the fossils we find imbedded in the rocks, as well as from the position of the rocks themselves, it is not difficult to geologically state both when the rocks were formed and when they were upheaved into mountain masses. A knowledge of rock-structure enables us to understand the agencies which have so altered its character as to produce the condition known as " metamorphic." We can readily see which rocks have been fused and molten from intense heat, and which have been literally *baked* and *stewed* at great depths below the surface of the present crust. We can see how once horizontally bedded rocks have been folded and squeezed, sometimes so intensely (as in the schists of Ben Lomond) that they look as if they had been crumpled like a handful of ribbons. Or we can see the strain to which some parts of the earth's crust entering into the structure of our hills and mountains have been put at various times, by the *dislocations* or "faults" to which they have eventually given way. Not unfrequently these slips extend to many thousands of feet difference in the continuity of the strata on each side such " faults," although it is often difficult to perceive from the surface arrangement of the masses that any such dislocation really occurs. This is due to the fact that denudation has long ago pared down and eaten away any surface disarrangement which might have been produced.

The amount of wear-and-tear to which our oldest hills and mountains have been exposed can hardly

be realized by any but those who have devoted themselves to the study of practical geology. Thus, we
know that from the top of Skiddaw there has been
removed no less a thickness than 30,000 feet of
solid rock! From the Western Highlands of Scotland, as well as from the surface of many Cornish
moors, strata from 30,000 to 50,000 feet thick have
been weathered away. It is certain that all the
now-exposed Laurentian and Cambrian rocks of the
north-west of Scotland were once buried beneath
thousands of feet of Silurian rocks. These have
since been eaten away, and the Cambrian and Laurentian rocks laid bare. The latter are the oldest
in the geological record, and such grand and abrupt
mountains as Suilven in Sutherland are entirely
composed of Laurentian rocks, surrounded by a
framework of hills formed of Cambrian strata, which
latter have been denuded from and about old
Suilven. The hills and mountains of Cumberland
and Wales are of later date, but they have hardly
suffered less from denudation. They are lower in
height now than ever they were before in the geological history of our planet, for their ridges and
summits have been continually weathered since they
were first upheaved above the general level of the
earth's crust. Most of our oldest hills are formed
of Cambrian and Silurian rocks. Nearly all those of
Scotland are, with the exception of the rugged mountains of Sutherland. The Highlands, the Pentland
and Lammermuir hills, the Grampians, the Cumberland mountains, those of Shropshire, North Wales,
of Connemara, and elsewhere are distinguished by

the large proportions of rocks of this particular age which enter into their composition. The softer strata of the later Old Red Sandstone period have been more eaten away, so that we never have rocks of this age forming high mountains, although they often compose undulating hills, such as those of Hereford-shire and the sides of the Mendips. In Devonshire the harder slates, formed later still, give rise to bolder scenic features, as in the Quantock hills and the neighbourhood of Dartmoor.

We next come, in the order of geological succes-sion, to such of our British hills and mountains as are formed of Carboniferous rocks. Mountains they seldom or never rise unto, for we rarely have eleva-tions formed by rocks of this age of more than 2,000 feet above the sea-level. They form hilly and moorland scenery rather than mountains. The grandest scenery produced among Carboniferous rocks lies where the lower subdivision of limestone comes up. Owing to its great hardness it enters into the structure of some of our highest and most extensive English hills, and in the earlier history of geology it went by the name of "Mountain Lime-stone" for this very reason. The finest hills formed of this rock are those of the Peak of Derbyshire. Penyghent, Ingleborough Fell, the Eglewysg Rocks at Llangollen, are other well-known hilly districts where this formation occurs. The Millstone Grit composes much of the hilly regions of Yorkshire, Lancashire, and Derbyshire, forming the Pennine chain ; Kinder Scout, in the latter county, being almost entirely composed of this rock. Then we come

to the strata of Permian and Triassic age, usually
soft and easily weathered, and therefore forming the
dales and plains of Cheshire, Worcestershire, and
elsewhere. Although occasionally we have some
member of the harder subdivisions rising into hills,
as, for instance, Alderley Edge, and Beeston Hill,
crowned by its grand old castle, in Cheshire. The
Oolitic rocks, which extend in a belt across England,
from the Yorkshire to the Dorsetshire coasts, frequently form hills and exhibit steep escarpments. The
Cleveland Hills and Cotteswold Hills are composed
of these rocks. The Chalk, in Yorkshire, Lincolnshire, Sussex, and Kent, rises into those rounded
hills termed "Wolds" and "Downs," which have
a scenery peculiarly their own.

Thus, it will be seen that our British mountains
and hills are formed of rocks of different geological
ages, and that both the hills and the materials composing them have their origins referred to various
geological epochs. It is believed that the Wrekin, in
Shropshire, is the oldest mountain in England, if not
one of the oldest in the world. Its main mass is
composed of volcanic materials, bedded and stratified,
which were ejected from a crater not far distant from
the present site of the mountain, during a geological
period long before the Cambrian. The base of the
Cambrian rocks in the neighbourhood is formed of
a conglomerate of pebbles actually derived from part
of the most ancient mass of the Wrekin. Deposition
has since then buried it beneath thousands of feet of
rocks of later date, and denudation, chiefly carried on
during the Tertiary period, has stripped them off again.

THE MOUNTAINS ABOVE GLENCOE, SCOTLAND.

Many of our British mountains owe large portions
of their masses to volcanic agency; although we
cannot say that many of them are ancient volcanoes. The highest mountain in the Isle of Mull is
above 3,000 feet high; and Professor Judd has
demonstrated that it is the boss or denuded base
of a volcano which was in active operation during such
a geologically recent period as the Miocene. When in
the zenith of its energy Professor Judd is of opinion it
must have been from 12 to 15 thousand feet in height.
The bosses of several ancient volcanoes in Cumberland
have been described by the Rev. J. Clifton Ward, some
of them, as Castle Head, forming notable scenic features
about Lake Derwentwater. Among those of our
British hilly regions indebted to volcanic agency for
the whole or part of their rock-masses may be mentioned the following :—The Cheviot range; the
mountains and hills in and about Kirkstone Pass, in
Cumberland, where there is abundance of green
volcanic ashes and volcanic breccias, in places 6,000
feet in thickness. The neighbourhood of Snowdon
(especially at Llanberis), of Dartmoor, the Wrekin,
the hills of Charnwood Forest, the Cornish hills,
Mount Sorrel, in Leicestershire; Rannock Moor, up
to which the famous Pass of Glencoe leads, in the
Western Highlands of Scotland, &c. These volcanic
results spread over many periods of geological time;
although there can be no question that the Silurian
was an epoch when volcanic activity was especially
great in what is now the British islands. Many
thousands of feet in thickness of rocky matter were
then disgorged, and turned inside out of the earth

through the mediumship of volcanoes. Here and there, as in the Cumberland and Welsh hills, we can hit upon the very places where this ejection took place untold ages ago !

It is doubtful, when the Welsh and Cumberland mountains were first formed, whether in the life-history of the globe, vertebrate types of animal life had as yet come into existence. For ages afterwards, during which the materials were being elaborated out of which other hills and mountains were to be sub-sequently formed, there were no warm-blooded animals living on the earth. Every chain of mountains charac-terises a distinct geological period, as M. de Beaumont and other geologists have shown, and we may therefore regard each elevation as a monument celebrating the introduction first of one and then of another of those great families of animals and plants, which geology teaches us succeeded each other in the vital history of our planet.

The Rev. J. Clifton-Ward, the best authority we have on Cumberland geology, in a paper recently published in the *Geological Magazine*, on "The Physical History of the English Lake District," gives the following account of the igneous origin of the rocks forming some of the Cumberland mountains. He says : "The rocks deposited in the present area of the Lake District during the period coming be-tween that of the Skiddaw Slates and that of the Upper Silurian are almost exclusively of volcanic origin. They may represent a total thickness of about 12,000 feet. At the base of the volcanic series only are there intermixtures with rocks of an ordinary

sedimentary character ; here, where the junction beds
are exposed, occur alternations of Skiddaw Slate and
submarine volcanic deposits. The rest of the series
consists of beds of volcanic ash and breccia with
lava-flows. The finer **ash** deposits are frequently
well stratified and false-bedded. The breccia is of
all degrees of coarseness, from a rock made up of
fragments having the size of a sixpence or shilling to
one containing blocks several yards in diameter.
Conglomeratic ash occurs in one or two beds *near
the base* of the series. Much of the variation in appearance among the beds of the ashy series is due to
subsequent alteration ; metamorphic action producing
diverse changes, dependent, oftentimes, upon slight
original differences in texture and composition—
examples of *selective metamorphism.* The lava-flows
are either good dolerites and basalts, or belong to a
class more or less mediate between these and the
more acidic group of lavas. As is generally the case
among volcanic deposits, the various beds are often
irregular in their range, showing instances of rapid
thickening and thinning.

   " The presence of ordinary sedimentary beds inter-
stratified among the volcanic deposits near their
base ; the occasional occurrence in this lower part
also of conglomeratic ash ; and the absence of both
these peculiarities in the great bulk of the volcanic
series, together with the absence of fossils in the
bedded ashes,—all point to volcanic action commencing at the close of the so-called Skiddaw Slate period
beneath the waters of the Skiddaw Slate sea, and the
gradual passage from submarine volcanic conditions

to those of terrestrial and wholly sub-aërial volcanoes.
At first sight it might seem that the regularly-bedded
ashes running at intervals throughout the series
pointed to subaqueous deposition, but no one can
ramble much around modern terrestrial volcanoes
without being struck by the frequent cases of fine
stratification shown by the ash scattered around,
whether deposited in the wet or dry state, and by the
not infrequent cases of false bedding.   It may sometimes have happened also that extensive deposits of
ashy material were laid down in large crater-lakes.

" The centres of eruption are difficult to fix upon,
as might be expected amongst volcanic remains of
such antiquity.   The boss of Castle Head, Keswick,
almost certainly represents one such centre, and the
best developments of lava-flows are all found occurring within an easy distance.   It may be further
remarked that since the lower part of the series contains the greatest thickness of lava-flows, it would
seem that the chief emissions of lava were followed
by long and continued ejections of ashy material.
What the height of the old Cumbrian volcano or volcanoes may have been it is difficult to estimate ; but
volcanic deposits were accumulated to a thickness, in
parts, of at least 12,000 feet ; and the highest beds
known (the fine, altered, almost flint-like ash of Great
End, Esk Pike, and Allen Crags) are unsucceeded by
any conformable series of sedimentary rocks ; hence
we know not how much of the products of the old
volcano has been lost, and, for aught we know to the
contrary, an Etna in size may have once stood where
now are the resting-places of quiet lakes.   In this

connection it is interesting to remember how little of our miniature mountain district would be uncovered, could we transplant Etna bodily with its surrounding volcanic ejecta to the site of the present Lake District."

Mr. Ward is of opinion that a great deal of denudation of these rocks took place during the Old Red Sandstone period, or the epoch between the Silurian and Carboniferous. Thus he says : " We come upon a period of time unrepresented by written records, but clearly evidenced by the destruction of the records of the previous periods. This destruction by denudation was carried on to an enormous extent, being accompanied, or rather rendered possible, by great movements of upheaval over the whole tract, which movements were probably most intense along a N.E. and S.W. axis, running through the heart of the present *mountain* district. If we pile up black cloth in layers to a thickness of ten inches, red cloth upon this to a thickness of twelve inches, and blue cloth upon the red fourteen inches thick, then by bringing some powerful force to bear upon the two ends of the pile, the whole may be thrown into curves and contortions by the lateral pressure, and the centre portions consequently raised above the level of the ends. Imagine, then, a large pair of shears brought forward which shall cut off or pare down all the upraised central portion ; in this way the uppermost blue cloth may be removed altogether from the centre of the low arch, a less amount of the red cloth layers would be removed, while perhaps only a few topmost inches of black cloth would be touched,

but the consequence would be that, at the centre of
the cloth dome, an arch of black cloth would appear,
on either side of this would lie inclined layers of red
cloth, to be flanked in their turn by similarly inclined
strata of blue cloth.   Suppose, now, the ten inches of
black cloth to represent 10,000 feet of Skiddaw Slates,
the twelve inches of red cloth 12,000 feet of Volcanic
deposits, and the fourteen inches of blue cloth 14,000
feet of Upper Silurian strata ; suppose, moreover, the
lateral pressure applied at either end of the cloth to
signify a probable depression of extensive tracts on
either side of that over which these formations are
upraised, and the gnawing and planing action of the
sea along the coast-line of rising land, aided by the
powers of the atmosphere, to be represented by the
great shears, then, while the elevation and crumpling
did their work, and were bringing the pile of 36,000
feet of strata conveniently under the denuding agents,
some of the outer coats of this dome must have been
pared away, and Upper Silurian and Volcanic deposits being both removed over the central dome,
the Skiddaw Slates themselves would be once more
exposed to view.   Such a denudation must mean
the removal of 20,000 to 25,000 feet of strata, unless
we suppose that within the distance of *a few miles*
(twelve or fourteen) the thickness of the Upper Silurian beds was much reduced.   To this amount of
removed material we must add a considerable thickness of Skiddaw Slates, themselves cut from the dome
top.   Surely such an action must represent a very
great length of time, yet do we find that it all tran-
spired in the interval between the close of the Upper

Silurian and the deposition of the Red Conglomerate of Mell Fell, ushering in the great Carboniferous period. For these Lower Carboniferous rocks (Mell Fell Conglomerate, etc.) are deposited across the denuded edges of all the older formations; at one place they lie upon Upper Silurian, at another upon rocks of the Volcanic Series, and at yet another upon the Skiddaw Slates. Thus there cannot be the shadow of a doubt as to the length of time which must have elapsed between the close of the Upper Silurian and the commencement of the Carboniferous period, and of the greatness of the work accomplished in that time.

"It is to the earlier part of this lengthy period, when the Skiddaw Slates were buried at their deepest, and internal commotions began to be displayed, that I would assign the formation of the various granitic centres. I have already treated of the probable pressure under which they were respectively formed, and called attention to the fact — in the Survey Memoir (p. 74)—that an axial line of the most intense metamorphism runs parallel with the main axis of upheaval. That there was a disposition at this period towards volcanic outburst I do not doubt, and that the mass of Shap Granite came nearest to establishing a volcanic connection with the surface I have hinted before ; but we have no evidence that the granitic roots were ever continuous, in this district, with volcanic vents, though, having no deposits of the true Old Red age (Lower Old Red) in the district, it would be unsafe positively to say that there were not volcanic eruptions within this area in Old

Red times, or that the Shap **Granite**, for example, does not represent the root of such an Old Red volcano. Of the three principal granitic centres the Skiddaw Granite occurs only in connection with the Skiddaw Slate, which is extensively metamorphosed around, but the granite, *where we now see it*, is undoubtedly intrusive ; the **Eskdale Granite** ranges for a distance of more than fourteen miles through the Volcanic Series, and is surrounded by an extensive zone of altered volcanic rocks ; while the Shap Granite has altered both volcanic rocks and the Coniston Series. We have, in fact, in these three masses, granite consolidated at various depths, or the *potential* roots of volcanoes exposed at different stages."

CHAPTER II.

THE ORIGIN OF MOUNTAIN AND HILL SCENERY.

Different Shapes of our Hills and Mountains—Mountains *not* upheaved into their Present Forms—Physical Geography of the Highlands—How the Highlands were formed—Sections across Mountain Ranges—Valleys produced by Denudation—Result of Denudation of Hard and Soft Rocks—Origin of Plains, Fens, Marshes, &c.—Valleys on Mountain Summits—"Cheese-wrings," how they were formed—*Débris* on Mountain Sides—How Passes, Gorges, and Ravines have been formed—Minor Scenery of Hilly Districts—Rounded Hills—Evidence of Former Ice-action—Evidence of Continental Ice Conditions—Ancient Glaciers among our British Mountains—Origin of Lakes, Tarns, &c.—Perched Blocks, Ancient Moraines, &c., of our Mountains—Eskers, Kaimes, &c.—Rev. J. Clifton-Ward on Glaciation of the Cumberland Mountains.

THIS part of our subject naturally grows out of that just discussed. To many people it may sound strange to hear the *origin* of scenery debated, for they have, perhaps, always entertained the idea that the present scenery of the globe was created just as we now see it. No mistake could be greater. Our physical scenery is the result of certain conditions and natural processes, all of which can be easily understood. The *different shapes* of our various hills and mountains, the gorges which seam their flanks, the passes which cross mountain-chains, the valleys which separate hill from hill, the tarns scattered among their quietest and most solemn recesses, and the peaceful lakes which repose at their feet and reflect their loftiest summits —all these characteristic features have been formed by certain operations, which it is the duty of the physical geologist and geographer to elucidate.

The first idea, in endeavouring to explain the origin of mountains, is to suppose they were upheaved into their present forms. There is just sufficient truth in this to lead a student wrong. It is true that all mountain-masses occupy areas of elevation, but it is not correct to conclude that the individual hills and mountains owe their shapes to mere elevation. Let us take the Highlands of Scotland as an example and illustration of the origin of mountains generally. If we could get a straight-edge long enough to extend across the entire region we should find it would nearly touch the tops of all the mountains it stretched across. We should then see the idea is here, at least, perfectly correct, that mountains owe their shapes entirely to the valleys which have been scooped out of them. This Highland region was once a lofty table-land, an elevated plain of marine denudation. In the long ages which have elapsed since the plain was elevated, the atmosphere has been wearing and tearing its surface. Rains, rills, and rivers have eaten into it, the softest rocks have weathered the quickest, and the hardest the slowest. Consequently, after this almost infinite wear-and-tear, we have now the present billowy arrangement of hills and mountains. The valleys occur, as a rule, where the rocks are softest, and the highest mountains are where the rocks of the once

highly elevated table-land were hardest.   Hence, as
Prof. Ramsay has well remarked, "Just as a railway
navigator leaves pillars of earth in a railway-cutting
to mark how much he has removed, so the great
excavator, Time, has left these mountain land-marks
to record the greatness of his operations."

*Section across a Mountain-Chain, showing the Highest Points occupied
by Hard Gneissose Rocks.*

Thus we are brought face to face with the wonderful fact that every mountain owes its existing shape
solely to the sculpturing agencies of the atmosphere.
Hence our oldest mountains are composed of the
hardest rocks, and of the latter we shall invariably
find the hardest of all, such as granite or gneiss,
forming the loftiest peaks and summits.   The common idea is, that if we were to cut a section across a

*Section across a Hilly Range, formed of the same Formation, showing
how Valleys are produced by Denudation.*

mountain-chain we should find the rocks lying against
each flank, and going over like a saddle, to form the
ridge.   This is very seldom, if ever, the case.   It is
true that strata have been often crumpled and folded
in the neighbourhood of mountains, for unless such a
process had taken place there would have been none

of the high ground of the elevated table-land for the weather agencies afterwards to operate upon, to eat into, and eventually to produce the present landscape effects by removing the softer strata, leaving the hard masses standing base to base as hills and mountains. This is sometimes the case when the same strata dip at a lower angle, for if we follow the same beds of rock over great distances we shall often find them relatively soft and hard. When the rock-masses were elevated we may be sure that atmospherical wear-and-tear would soon find out the weakest and easiest places to operate upon, and there, in the long

Section across a Hilly Range, showing Strata of same angle of dip, but of different degrees of hardness, the hardest rising highest.

run, we should have hollows scooped or eaten out. The drainage of the surface water would naturally be along the excavated places, and so would assist in still further deepening them, until eventually valleys would be formed, in some of which rivers would meander. When a formation had its beds upheaved from their horizontal position, it would be foolish, indeed, to expect that all the strata would have just the same degree of hardness over a large tract of country. If they had not, then we may reasonably expect to find the softest of them eaten away the most, so that the outcrops of the hard rocks would take place on the loftiest parts of the hilly range.

Every kind of rock has its own style of weathering,

and a skilful geologist can actually tell a geological
formation by the scenery of its surface, by the shape
of its hills, and the contour of its valleys. Chalk is
usually marked by the round-topped hills we call
" Downs ;" granite weathers into elevated rolling
moorland ; gneiss, mica-schist, and metamorphic rocks
generally form the most inaccessible mountains,
having needle-shaped peaks ; carboniferous limestone
is usually characterised by a terrace-like arrangement,
according to the dip of its strata, and by the presence
of numerous gorges and narrow " dales," like those
of Derbyshire and Yorkshire. The Yoredale Rocks,
the Lias Shales, the Oxford Clay, Kimmeridge Clay,
and London Clay, are usually met with in large valleys,
plains, and fens, because the softness of these strata
has caused them to be weathered to the lowest level,
and, as they hold surface water, they usually produce
fenny or boggy areas. The " fens " of Cambridge-
shire, Norfolk, and Lincolnshire, and the " marshes "
of Essex, are all underlaid by one or other of the
above-mentioned soft clayey formations.

In the saddle-shaped (*anticlinal*) and basin-shaped
(*synclinal*) curves into which the strata of a district
have been bent, it might be thought that the former
produced mountains and the latter valleys. Such an
assumption, however, would be totally without proof.
The shapes of our hills and mountains are altogether
independent of the flexures into which the rocks
composing them have been thrown. Synclinal or
basin-shaped depressions of strata are quite as likely
to be found on mountain-tops as in the valleys beneath. In not a few instances the summits of our.

hills contain fragments of ancient valleys, as is the case with the Scuir of Eigg.

It is more than probable that all our oldest British mountains stand at a less altitude now than they have done for millions of years past. For during that time they have been exposed to various weather-action, which has stripped or peeled off large quantities of rocky material even from the hardest of them. Meantime the valleys have been gradually widening, and eating farther into the heart of the hills ; and not unfrequently we find instances where the valleys have been thus carried back from two opposite directions, until there is only a thin partition of rock-barrier now separating them. Given time enough and a continuance of the same weathering action, and eventually even this barrier would be eaten away, and the two valleys be thrown at last into one. There can be little question that this mode of forming one large valley out of two smaller ones has often taken place in the physical history of our mountains.

How even the hardest and ruggedest of rocks suffer from the action of the weather is curiously illustrated in many of the hilly districts of Great Britain. Along the crests of some ranges of hills we may see piles of stones, one upon another, to which the names of " Cheese-wrings," " Cakes-of-bread," &c., are popularly applied. We may see them to perfection along the crest of the Millstone-Grit which is crossed on the road from Sheffield to Castleton, in Derbyshire. The lower stones are frequently of narrower diameter than the upper, owing to the

mechanical action of the dust wearing away the
base.   These stones, piled on each other like
cheeses, are simply so many surviving fragments of
once-continuous strata, all of which have gone except these miniature "outliers," existing to indicate
the amount of wear-and-tear which has taken place
even since the conclusion of the "Glacial period."
An observer may easily perceive, by noticing how
these piled stones are more or less continuous in
their arrangement, that they represent strata of solid
rock which have been weathered away to at least
a thickness amounting to the highest point of the
"cheese-wrings."   Not unfrequently these rounded
stones are so delicately poised that they have been,
and still are, regarded with superstitious awe by the
unlearned.   They are looked upon as the work of
demons, fairies, or oftener still, of Druids. The name
of "Druids' stones" is frequently given to them,
especially in Cornwall.   These weathered blocks are
of commoner occurrence in Palæozoic rocks than any
other.   In Granitic districts, as on Dartmoor, we may
often see rounded piles of granite, like so many
heaps of boulders, each rounded mass being
weathered everywhere except where it is protected
by the surface of the overlying stone. Sometimes
great masses weather, *in situ*, in such a manner all
over them, that the base rests on a point so small
that the block can be easily rocked to-and-fro. Such
rounded, weathered, and delicately-poised blocks are
called "Logans," or "Rocking-stones." In Cornwall
these have even been called by *individual* names, as
the "Nut-cracker," in Lustleigh Clewe, Devonshire.

One of the best known is situated near Castle Treryn,. St. Leven, Cornwall. Its size is about 17 feet in length, and more than 32 in the circumference of its middle part; the total weight being calculated at about sixty-five tons. A "Cheese-wring" which occupies the highest ridge of a hill north of Liskeard, in Cornwall, is piled up to the height of 17 feet. Most people who have visited Dartmoor,. Torquay and Newton-Abbot will vividly bear in mind the conspicuous appearance of another of these evidences of atmospherical waste known in the neighbourhood as "Rippon Tors."

Next to these evidences of weather action (which are perhaps the most noticeable because they often stand out sharply against the sky) is the *débris* which may be seen on most hill and mountain sides. We know that, for the most part, this talus must have accumulated since the close of the Glacial period. There is scarcely a hilly slope which is not covered with blocks that have been loosened and fallen down to where we now see them. The larger of these blocks after they have fallen undergo weather-action in the places we see them occupying, and are there further subdivided, until eventually they are reduced to unnoticeable fragments. The surface of the hill on the left-hand side as we descend Kirkstone Pass towards Brothers-water is literally covered with huge fragments of rock, one of them so large that on account of its resemblance to a simple chapel, or kirk, it has given its name to the pass. Thereabouts, these boulders (as they are mis-named, for they are mostly detached weathered fragments from the crests above)

are set in the most delicious emerald framework of
Parsley-ferns. The slopes of the Millstone-Grit hills
of Lancashire and Derbyshire—where they allow of
the detached rocks coming to rest on their surfaces,
and are not too steep to force them to roll to the base
before they can stop—are often delightful paradises
of ferns and other rare plants, which nestle beneath
the shelter of these reposing stones. A wildness,
truly, does a hill-slope like this produce, of gnarled
saplings, angular moss-grown stones, lichen-clad boulders, damp, oozing-moss tracts, out of which emerge
sparkling mountain streams, with banks verdant with
liver-wort and jungermannias.

This old crop of rolled fragments gradually weathers
away, and the mountain streams carry off the detritus
in the shape of large sand-grains. But a new crop
succeeds the old; and in this manner, through the
silent but unresting centuries, the highest mountains
are brought low, whilst the valleys are being exalted.

Let us turn to one other special agency we can
see at work sculpturing the mountains and hills
into characteristic shapes, or giving to them some of
their best-known features. The rills and waterfalls
which carry off the drainage of the heavier rain-falls
that usually visit mountainous regions are wonderful
natural implements for sawing their rocky beds backwards ; and it requires little special training for the
student to see them at work in every stage and degree
of development along every mountain flank. Possibly,
during some walking tour in North Wales or the
Lake District, the evening has been inaugurated by
a steady downpour of rain, which has continued

persistently all night, and only ceased the following morning. Then the morning sun has broken forth with more than summer glory. The hills and meadows are bespangled with raindrops, each of which reflects the sunshine. The brooks and rivers have risen in turbulent volume, and continued to be swollen

WATERFALL CUTTING A GORGE OR RAVINE.

all the day by the contributions of the mountain torrents. Every mountain side is seamed with what appear to be silvery threads, but which we know are the improvised rills carrying off the excessive rain-fall. They only appear on occasions like these. Here and there, perchance, is a more regular water-

course, in whose bed you discover a miniature rill
even on the hottest and dryest day in summer. It is
the home of the dipper, or water-ousel, who finds
abundance of insect-life among the moss-clad stones
which abound in such a gorge. Possibly you may
see defiles deeper and larger still—in short, we get
them of all "sizes," from the gutter-like seam which
marks the presence of a temporary rill in rainy
weather, to deep and even awful clefts, along whose
bottom there is one perpetual roar of running water.

Examine any or all of these various-sized indenta-
tions in the mountain or hill-side, and you will invariably see their beds bouldery and rocky, and full
of all sorts of loose fragments of stone. We have
said that anyone might find an almost imperceptible
gradation in the various rills, between the smallest
and most insignificant, and the deepest and gloomiest
gorge. The origin of the latter now suggests itself.
We can easily see that after a heavy rainfall, when
the descending waters are rushing down the water-
courses, all the boulder and rock fragments lying
along the bed will be set in motion. Their mechanical action wears the channel deeper and wider, and
when this agency has been in operation for un-
numbered thousands of years, that which was once a
mere indentation down where a silvery mountain
stream found its way after a storm, will have become
a "burn," a "glen," or a "gorge." In this way
there can be little doubt that all such wild clifts
or bonny mountain glens have been mechanically
excavated by the action of running water. We have
seen how ancient our British mountains are geologi-

cally, and we have therefore no difficulty in giving to
agents still in operation an abundance of time for them
to eat out the most wonderful of our mountain gorges,
glens, and even such passes as those of Glencoe,
Llanberis, and Aberglaslyn.   So awe-inspiring are
some of the rifts which thus appear in the solid rocks
that even intelligent people feel bound to explain
them by reference to earthquake shocks.   But the
beds of the streams show that the rocks are continuous
across.   There is no fissure or cleft such as would
be produced if the defile had been rent open.   Every
geological observation leads to the conclusion that this
characteristic kind of mountain scenery has been due
to the "small, still voice" of rills and streams, silently
but perpetually operating during innumerable ages !
Every waterfall and cataract wears backwards the ledge
over which its waters are precipitated, and thus, like
a vertical saw, each gorge is slowly eaten backwards
into the heart of the hills.

The *minor* glen-scenery of hilly districts has very
probably originated since the close of the Glacial
period.   That was the last of the great and important
geological epochs, and it immediately preceded the
appearance of Man upon the face of the earth.   It was
a most important period in many respects, but in none
more than the way in which it gave the last finishing
touches to the scenery of our hills and mountains.   As
the name given to it by geologists indicates, it was an
age peculiar for the extreme cold which then prevailed over the whole northern hemisphere.   The
probable origin of this cold period is not unknown to
astronomers and geologists.   Suffice it for us at

present that there is abundant evidence to prove it
continued in force for many thousands of years, with
intervals of comparatively warmer conditions.   At the

ENTRANCE TO THE PASS OF ABERGLASLYN, NORTH WALES.

earlier part of the Glacial period our British hills and
mountains were loftier than they are now, for they
suffered subsequently severely by glacial denudation.

Could we have obtained a glimpse of our mountain scenery before this period of long-continued cold set in, although we should have beheld a good deal of the main features of the landscapes which still impress us, we should have been more impressed by the *absence* of certain phenomena, which are now very striking in such districts, rather than otherwise. It is more than probable we should not have seen any of the lovely lakes which now so charmingly diversify the mountain scenery of North Wales, Cumberland, Scotland, and even of Ireland. Mountain tarns—those quiet and solemnly suggestive spots of which Wordsworth speaks—there were none. Perhaps all these picturesque aids to the loveliness of our mountain scenery are due to the causes which were in operation during the Glacial epoch.

Let the tourist only examine the general contour of many of the lower hills in our mountain districts. Although they are formed of rocks of various geological periods, and even of different mineralogical compositions, he will observe how they all alike assume a rounded and flowing outline. Towards their bases the lesser hills are particularly carved in this way. How is this? To the trained eye of the geologist there is only one agent in nature capable of producing this uniform shape of hill and mountain — ice! He is aware that the rocks near the existing glaciers of the Alps have been rounded and smoothed by ice in a similar way. He knows that if our British mountains were once again to be swathed in a slowly accumulating and moving ice-sheet, the mechanical pressure of the continuously

moving mass would wear and tear the underlying
rocks, and round off the sharp edges, until one
general feature had been impressed upon all the
rock-surfaces alike. The tourist can verify this
glacial interpretation of the abundance of the rounded
outlines of the lower hills for himself. As he wanders
amid the mountain solitudes he may frequently see
the *ancient ice-marks* caused by the ice sheets of the
Glacial period. Every hill and mountain in Great
Britain (if its rocks are hard enough to retain the
impressions) is scratched, scored, or polished by ice
action. In many places you have only to peel off
the thin layer of turfy sods for your eyes to gaze
upon the ancient ice-markings of this wonderful
period. Darwin graphically expresses one of the
best-substantiated of geological facts when he says:
"The ruins of a house burnt by fire do not tell their
tale more plainly than do the mountains of Scotland
and Wales, with their scored flanks, polished surfaces,
and perched boulders, of the icy streams with which
their valleys were lately filled." The source and
direction of many of these ancient glaciers has been
tracked and mapped; and in the maps, for instance,
accompanying the Rev. J. Clifton-Ward's "Memoir
of the Geology of Cumberland," he will see the
centres of dispersal, and the courses taken by the
dispersed and moving ice-sheets. The flanks of the
hills and the sides of the valleys are frequently well
marked by the stones and rock-fragments which were
imbedded in the ancient glaciers that scratched
and grooved them. High up the passes of Llanberis
and Aberglaslyn, in North Wales, the young geologist

may still see the lines of grooving left by the glacier which once passed down these ancient defiles into the valleys below. In the former pass the marks are in places nearly as evident as those left by the ropes of a canal-boat upon the edges of the brick-work of the bridges against which they have repeatedly grazed.

In the earlier part of the Glacial period the cold reached its maximum. Eventually all the high hills and mountains were covered with and masked by ice. Could we have seen Great Britain at that time it would have appeared in the physical condition in which Greenland is now. The latter country is utterly unknown, except a strip on the western side where the warm waters of the Gulf-stream keep back the ice. Everywhere else, for a space of twelve hundred miles from north to south, and of four hundred miles from east to west, Greenland is completely clad in a densely-thick and slowly-moving sheet of ice. No man has travelled further than thirty miles inland upon this snowy wilderness. All the moisture that descends falls as snow ; and unless the excess were removed by local glaciers, there would be built up eventually a huge pyramid of snow and ice. The ice-wall moving seawards is in places hundreds of feet high, and fragments detached by it as it moves, or wears away by the breakers of the wintry sea, float away as icebergs. This condition of Greenland is known as that of *continental* ice, as distinguished from the local streams of moving ice proceeding from the snow-line of the Alps, down some valley, to a level low enough for the increased temperature to

melt and check the ice-flow, and to convert its substance into water. Such mighty rivers as the Rhine
and the Rhone have their origin in the constantly-
melted ice and snows of Alpine glaciers.

Our British mountains are stored with abundant
traces of a former *continental* or Greenland-like condition of ice. Both in Scotland, Cumberland, Lancashire, and Yorkshire it has been demonstrated that
the land-ice was thick enough to fill up the valleys,
as well as hills and even mountains, for the path
of the continental ice-sheet is still plainly traceable.
It was in this way that some of our lower mountains
and so many of our hills received that general
rounded outline which they at present exhibit. How
much they must have been denuded by this glacial
action is proved from the great sheets of *till* or
"boulder-clay," which are strewn over the plains of
Yorkshire, Cheshire, Lancashire, Nottinghamshire,
Leicestershire, and many other counties, but which
so densely cover almost the entire surface of Norfolk
and Suffolk as to cause the celebrated agriculture of
those places to be indebted to its presence. This
sheet of "drift" frequently attains hundreds of
feet in thickness, and covers thousands of square
miles.

It was at the latter part of the Glacial period, when
the continental ice-sheet had been stripped off, that
glaciers filled our mountain valleys. Many of these
valleys had been filled up with drift rubbish, which
the glaciers now ploughed out. In this way many
ancient valleys and gorges were re-excavated, and
as such we now behold them. It was these local

glaciers which gave the last finishing touches to our mountain scenery. Some of them widened the gorges and passes and glens. Where the previous ice-sheet had pressed down in the greatest mechanical force along the bottoms of old valleys, the solid rocks were slowly denuded and worn off, until at length a "rock-basin" was formed. These "rock-basins" are not due to a local depression of strata, but to an excavation out of the solid rock of a hollow to a greater or less depth. We only find them in the neighbourhood of mountains, and of mountains which yield abundant traces of ice-action. When the Glacial period passed away, and the existing state of things was ushered in, the ice-sheets and glaciers disappeared before the increasing warmth. The "rock-basins" great and small were now the natural drainage places for mountain streams and rills, and so became the lovely lochs, lakes, and tarns which beautify the mountain districts of North Wales, Cumberland, and Scotland.

The glaciers left rubbish-heaps at the mouths of the valleys, whose upper parts they had filled with ice. These rubbish-heaps are "moraines," full of angular, scratched stones, just as we may see them in the recently-formed "moraines" of existing Swiss glaciers. These rubbish-heaps are abundant along the low-lying shores of many of our lakes, such as Windermere, Derwentwater, Coniston, and in North Wales; and much of the softer landscape, richly tree-clad, of these localities, owes its features to their presence. In North Wales occasionally a moraine-heap dams up the mouth of a valley, or occurs as a ridge across it, and thus forms a lake or large tarn. In the north of England, and particularly in Scotland and central Ireland, the glacial rubbish-heaps have been sorted and re-arranged by the waves of the sea since they were originally formed ; for we know that the larger area of Great Britain was submerged to as much as 1,200 feet at the close of the Glacial period. The names of " Kaimes " and " Eskers " are given to these re-arranged rubbish-heaps. They frequently form very wild and weird-looking scenery, for the rubbish-heaps wind like serpents over the valleys and plains, or at the feet of the hills ; and in Ireland and Scotland superstitious tradition has many a story to tell of the origin of such peculiar features in the landscape.

How softly and gently the ice left the hilly regions it once covered is indicated by the millions of loose stones or boulders, some of them of very great size, which are strewn over the surfaces of the gentler slopes of the mountains, or on the heights of the lower hills beneath. There they occur as " perched blocks," sometimes so delicately poised that they can be easily moved by the hand. If you examine them you see these boulders are usually formed of a different material to the rocks on which they rest. They are frequently perched in the most grotesque and impossible of positions ; and it needs little observation to perceive that these boulders could not have been detached from above, and rolled down to their present positions, for such an impetus would inevitably have carried them below, where even now it seems as if a child could push them. But when we know that ice-sheets once covered all the ground where these boulders are now seen, and that the moving ice-sheet must have had its surface strewn with detached blocks, which would be quietly let down when the ice melted, we have at once an explanation of the peculiar carrying agent which has placed them in their present positions. The low-lying parts and rounded terraces of such places as the Pass of Llanberis are crowded with perched blocks, some of them reposing in the most precarious situations. They are noticeable objects throughout all the hill and mountain scenery of Great Britain. Wordsworth refers to them, as he does in his " Prelude " and " Excursion " to many other features in the land-scape of the Lake district, which the geologist knows were produced mainly by the powerful ice-tools in operation during the period we have re-ferred to,—

" As a huge stone is sometimes seen to lie
Couched on the bald top of an eminence,
Wonder to all who do the same espy.
By what means it could thither come, and whence,
So that it seems a thing endued with sense,
Like a sea-beast crawled forth, that on a shelf
Of rock or sand reposeth—there to sun itself."

The amount of wear-and-tear due to weather action, which has taken place since the mechanical action of ice gave its last finishing touches to the scenery of our landscapes, is best seen in the "Cheese-wrings," &c., of Derbyshire. The reason these are so much more abundant in Devon and Cornwall than among the mountains of Wales, Cumberland, Scotland, and

Ireland, is because Devon and Cornwall were left uninfluenced by the ice-sheet which in the other localities was so active. We have little or no evidence of glacial action in South Devonshire and Cornwall; neither is there any proof of those areas having been submerged when North Britain was last under the sea. The consequence is that the hills of Cornwall and Devonshire have been longer exposed to the kind of weather-action that is now going on than any parts of England, and the results are seen in those piles of rounded stones which represent the slow weathering that has removed from extensive areas the strata of which they are the dwindled representatives.

The Rev. J. Clifton-Ward, in his "Geology of Cumberland," recently published as one of the "Memoriæ" of the Geological Survey, gives an outline of the conditions which affected the Cumberland mountains during the Glacial epoch as follows:—"At the commencement of the cold period small glaciers occupied the heads of the various valleys, and as the cold continued and increased they became larger and larger, until, in many cases, they united, overlapping the lower ridges parting valley from valley, and forming one great confluent ice-sheet, the movement of which was determined to the north and to the south, or east and west—as the case might be, in different parts of the district—by the main waterparting lines. A great quantity of rocky *débris* was moved onwards and left scattered over the country, partly by the first-formed moraines being pushed forward, and partly by the ice overriding the same and dragging on and triturating the fragments be-

neath it. In this way the Till was formed, sometimes
left in rock-sheltered upland hollows, but most largely
deposited on the lower and less inclined ground.

"Whether this first land-glaciation was interrupted by one or more mild periods, the deposits in
this district do not prove. As the final close of this
epoch of intense glaciation drew on, moraines were
left by the retreating glaciers plentifully scattered in
every valley, but the glacial streams and rivers must
have made much havoc amongst them, cutting them
up and bearing away their material to lower levels.

"Then, when the cold had disappeared, began a
submergence of the district to a very considerable
extent. As the land sank, the old moraine material
was sifted, sorted, and partly rounded. At the ends
of some of the fiords or straits, sand-bars were formed;
but, as there was no floating-ice during the earlier
stages of submergence, these sand and gravel deposits
enclosed no large boulders. The district became
gradually converted into an archipelago, and currents
circulated among the islands. When depression had
gone on to the amount of 1,000 feet or less, the cold
returned, and ice-rafts bore blocks from one part to
another. In many cases the direction in which currents swept the floating-ice was the same as that of
the old glaciers, and thus boulders were transported
along the same course at different periods and by
different means. Sometimes, however, or at certain
parts, marine currents bore floating ice, with its
boulders, in directions opposed to, or much at vari-
ance with, the old glacier courses. Thus, when the
land stood about 1,200 feet lower than at present, a

current, sweeping the north-western outskirts of the district, carried boulders from Sale Fell southwards on to Broom Fell. Not until the submergence reached over 1,500 feet was there any *direct* communication between the northern and southern halves of the Lake District, *except* by the straits of Dunmail Raise. Under such conditions a current very probably ran through those straits from south to north, turning mainly to the east on reaching Keswick Vale, though probably sending a branch off to the west ; while other currents may have set through the straits between Skiddaw and Blencathra. The case mentioned of an ash boulder at the upper sources of the Caldew would seem to point to a current having *at one time* passed from south to north, *up* the Glenderaterra Valley and *down* through that of the Caldew. The block could not have reached its present situation from any of the volcanic deposits lying north of Carrock and Comb Height, and is scarcely likely to have come from the very limited *ash* exposures of Eycott Hill, to the south-east of the Caldew at Mosedale. It is not likely that this boulder could have been transported by glacier-ice or any form of ice-sheet, because it is in the very midst of lofty mountains which would have produced sufficient ice to have filled the valleys between them and kept out any ice-sheet foreign to this group. Hence, I am inclined to consider this case as a *proof* of submergence to the height of at least 1,300 feet, and of the existence of marine currents passing through the Skiddaw mountain group.

"The submergence continued until the land must

have sunk more than 2,000 feet below its present level, as the position of boulders in many parts of the district seem to show, and notably those on Starling Dodd. Then the whole district was represented merely by scattered islands clad in snow and ice, each a little nursery of icebergs.

"As the land was re-elevated, the glaciers crept down to the level of the sea, sometimes forming moraines just at the sea-margin, as was the case beneath Wolf-Crag, Matterdale Common, when the land stood 1,400 feet below its present level. During all this time numerous boulders were let fall upon the early-formed mounds of sand and gravel, and as the sea shallowed, more of such mounds were deposited, now, however, frequently *containing* ice-borne blocks.

" Finally, when the land had regained its former height above the sea, glaciers still lingered in the recesses of the mountains, but this second set of glaciers at no time equalled the first in size. Some, in Borrowdale, were sufficiently large to creep down probably as far as Rosthwaite, and the more or less perfect moraines in every upland valley remain as the last traces of the Glacial period.

"I think too that the immediate cause of the numerous lake-hollows, *many* of which are now completely filled with stream-borne detritus, was the onward movement of the glacier-ice when at its thickest, as suggested by Professor Ramsay's theory ; for not only are these lake-basins extremely shallow when compared with the heights of the mountains and the thickness of the former ice-sheet, but in most cases the agreement is remarkable between the spots at which the greatest depth of water occurs and those points where, from the confluence of several ice-streams or from a narrowing of the valley, the onward pressure of the ice must have been greatest.

"I have sometimes been asked whether the glaciation described in my papers on this district was not wholly belonging to the close of the Glacial period, and whether previously the whole mountain district had not been overridden by a great ice-sheet from the north? To this I answer that, so far as I am able to speak, there is *no* evidence in the district of such a mountain-and-valley-ignoring sheet, and not *one* boulder of foreign northern rocks to be found *among* the mountains."

CHAPTER III.

THE SCENERY OF OUR HILLS AND MOORLANDS.

Position of our Moorlands—Origin of Soils—Bog-lands and
"Mosses"—Chat-Moss—Boulder-clay and the formation of
Bogs and Mosses—Climatal Effects of Draining Mosses, &c.
—Scenery of the Moorlands—Heaths—Moorlands of York-
shire and Lancashire—Ancient Forests imbedded in Moor-
land Peat—Rossendale Forest—Boulders on the Moors—
Rannock Moor—Moorland Streams—"Pot-holes," "Swal-
low-holes," &c., on the Yorkshire Limestone Moors—The
Peak of Derbyshire and its Valleys—Mam Tor, or the
"Shivering Mountain"—Barony of the Burren—How
Gorges are formed in Limestones—The Cheddar Cliffs,
Winnats, and the Derbyshire "Dales"—Eldon Hole—
Dove-holes—Messrs. Lees and Davis on Pot-holes, &c., in
West Yorkshire—Mr. Kinahan on Subterranean Rivers in
West of Ireland.

THE term "moor" as applied to the uncultivable
lands of the midland and northern counties and of
Scotland is derived from the Anglo-Saxon word
môr. It refers to any area necessarily laid waste on
account of excess of water or chaotic confusion and
diffusion of rock fragments and boulders, and the
riotous exuberance of the growth of heather and
gorse. Most of these moorland tracts are underlaid
by very poor soils ; often in Yorkshire and Lancashire
the latter are composed of nothing more than the loose
and large sand-grains produced by the decomposition

of the millstone-grit rock beneath; or, as on Rannock
Moor, by the slow rotting of the felspar of the
granite. Many of these tracts occupy high elevations, as along the broad, denuded crest of a range of
hills or the comparatively flat spaces which one
meets with in the heart of the hills. Or they may
stretch over the fragmentary surfaces of the weathered
and denuded table-lands out of which so many of
our hills and mountains have been slowly carved by
meteorological agencies.

In the low-lying moor-lands a good deal of space,
formerly deemed "waste," has been regained by
deep-draining and other methods of scientific agriculture. These lower wastes go by different names
in different parts of Great Britain. In Ireland
they are the well-known "bog-lands," in the eastern

Geological Section, showing the Position usually occupied by the
Moorlands of Yorkshire and Lancashire.

counties of England the "fens" and "marshes," in
Lancashire and Yorkshire they are called "mosses"
—a name which is also given to them along the
borders. The most famous of the latter is that
known as Chat-Moss, across which the railway from
Manchester to Liverpool is carried, which latter had
almost to be given up in the construction on account
of the seemingly inexhaustible capacity for rubbish
which the soft and boggy "moss" possessed. Since
that railway was formed more than half the Moss has

been reclaimed, and the traveller now sees waving fields of corn where only a few years ago all was watery morass. This area has remained in the same boggy state probably ever since the close of the Glacial period, for the moss is underlaid by a sheet of impervious boulder-clay, which retains the moisture, and has allowed of such a long-continued and

CHAT-MOSS.

dense growth of aquatic and moisture-loving vegetation that at length a thick layer of turf or peat has been formed, like the peat-bogs of Ireland, many of which originated in a similar way, and are indirectly due to the boulder-clays.

The sour, black soils of the "mosses" have been artificially mixed with sand and lime, and their quality is

improving year by year. In the fen country, as well as
where the "bogs" and "mosses" have been drained
and cleared and otherwise prepared as arable land, the
crops raised even now are unequalled in any other
parts of England. Moreover, the drainage of such
large areas has affected the climate generally for the
better. The heat of the summer's sun, which was
formerly expended in evaporating the water, and rendering our atmosphere more foggy than it otherwise
need be, is now directly exercised in heating the surfaces of the drained soils, and in warming the air
immediately resting upon them. The mean temperature of Bolton, in Lancashire (formerly known
as Bolton-le-Moors, on account of the dreary extension
of waste lands around it) has been raised more than
ten degrees since the moors were drained. This will
serve as an example of the beneficial effects of drainage.
The scenery of the unreclaimed tracts, wherever seen
in Britain, is wonderfully alike. There are the dense
clusters of bulrush and water - flags marking the
presence of the sloppier tracts, and clusters of dingy
alder grouped around dark, lonely pools, whilst the
general surface is adorned with a besprinkling of
dwarfed white birch-trees, with their silvery bark and
graceful, drooping foliage. In April and May the
entire area is aglow with masses of bright yellow
marsh - marigolds and buttercups, and by and by
chequered with the snowy abundance of the cotton-
grass.

But it is with our upland wastes, our moorlands
proper, that we have most to do. Who that has
wandered over them has not been impressed with the

sense of physical liberty which the scents of the heather and the gorse seem to bring with them? The far-stretches of billowy moorland and distant horizon, the breezy feel of the moorland wind, the very cries of the curlew and the plover seem to be wilder and less conventional than those of other birds —all these, and a score more unutterable natural suggestions of unbounded freedom, make "a day on the moors" an event to be treasured and remembered! No wonder that men who spend the greater part of the year in town, in the weary work of law-courts, parliamentary drudgery, or mercantile excitement, should expend their wealth in the preservation of a northern "moor," or that the sport they most affect should be grouse-shooting. In nine cases out of ten it is not the love of sport which impels them to this, but the unexpressed sense of mental and physical liberty which days spent on the moorlands invariably bring with them!

Edwin Waugh, the Lancashire poet, many of whose songs are full of what Wordsworth calls the "power of hills," thus speaks of the moors which extend over so extensive an area in East and North-East Lancashire and the adjoining West of Yorkshire—

"Sing, hey for the moorlands, wild, lonely, and stern,
  Where the moss creepeth softly all under the fern;
  Where the heather-flower sweetens the lone highland lea,
  And the mountain winds whistle so fresh and so free!
  I've wandered o'er landscapes embroidered with flowers,
  The richest, the rarest, in greenest of bowers,
  Where the 'throstle's' sweet vesper at summer day's close,
  Shook the coronal dews on the rim of the rose;

But, oh for the hills where the heather-cock springs
From his nest in the bracken, with dew on his wings!
I've lingered by streamlets that water green plains,
I've mused in the sunlight of shady old lanes,
Where the mild breath of evening came sweetly and slow
From green rocks where bluebells and primroses grow;
But, oh the wild hills that look up at the skies,
Where the green brackens wave to the wind as it flies!"

Perhaps the botanist and naturalist enjoy such scenery even more than the sportsman, and that without a fraction of the expense. To them every living object is a silently eloquent exposition of nature and nature's God! The yellow and grey lichens which crust the rocky fragments, the club-mosses whose ropes trail along the surface of the springy turf, the tiny wild flowers which nestle beneath the shadowing protection of the boulders year after year, the humming insects, whose presence gives you a delicious sense of loneliness—these are a few of the surroundings of a northern moor. The faint, but delightful suspicion of peat-smoke which every now and then greets the nostril, suggests the presence of a few human settlers; even their thatched, grass-covered cottages add to the sense of change which forms part of one's personal enjoyment. And then the pedestrian, footing it over the unequal surface of the moor, having to skip here from boulder to boulder to avoid the suspicious quagmires, there carefully treading on to a too-springy green piece of bog, now rushing through groves of bracken, mountain-fern, and blechnum, startled by and startling an unheeding covey of grouse, or a voracious sparrow-hawk, disturbing the dragon-flies which haunt the bubbling, sedge-margined pools, or temporarily deceived by the clever feints of the lapwing to draw him from the neighbourhood of its nest—finds the fleeting hours even of a long summer day all too short for him !

Or maybe the moor partakes of the character which gives such waste grounds in the south of England the name of "Heath." The soil is *dry*, not moppy nor spongy, nor underlaid by a thick layer of peat, as many of the northern moors are. In May such tracts are all ablaze with the yellow blossoms of the gorse (or whin or furze, as this shrub is variously called in different parts of England). Surely no sight is more lovely ; and if the story be not *true* that when Linnæus, the great botanist, first saw an English gorse-bush in full bloom, he knelt down and thanked Heaven for a sight so beautiful, it ought to be ! The tender young shoots of these bushes, ere they are hardened in the summer into the protective spine which keeps susceptible skins at a safe distance, are capital feed for mountain sheep. The young rabbits which usually abound in the same localities, as the gorse (for the dry soil suits both animal and plant) love to browse upon the succulent and newly-grown stems. As the summer advances it may be that a new object will make its appearance. Dodder-plants will shoot up from the ground and interweave their winding, thread-like stems in and out even of the spiky and prickly gorse, so that for hundreds of square yards all the bushes

will be in an indescribable entanglement. Setting
the scattered boulders in framework of purple-heather,
we shall find three or four species of that lovely plant,
and its ally the common ling. As the summer pro-
gresses the bossy uplands which catch the sunlight
will be lit up with the brilliant colour produced by
the massing together of the heather. Around and
about them the atmosphere will be fragrant with
their odorous breath, and musical with the hum of
insects engaged in seeking out their nectar. The
purling of miniature streams, more than half-hidden
in the groves of fern and whinberry amidst which
they wander, add to the charm of the scene. The
naked, thinly-grassed turf is begemmed by scattered
hosts of blue hare-bell and yellow mouse-ear and
hawkweed. The moister places are mapped out by
patches of purple louse-wort and willow-herb, or by
greener and denser groves of grasses unseen elsewhere.

Higher up amid the hills and along the shoulders
of the lower mountains, where our moors extend in
all their silent impressiveness, the wandering botanist
finds many a species of flower unknown in the valleys
beneath. Of these we shall have to speak at greater
length presently. But the distribution of moorland plants is largely affected by the character of
the rocks on which they grow. Thus, commencing
near York, we find that the western moorlands are
formed over the interrupted plateau of the Millstone-
Grit formation. In its western outcrop this well-
known and easily-recognised rock has been in places
much denuded, so that extensive valleys have been

literally excavated out of it. On either side these usually fertile valleys or "dales" (Wharfedale, especially near Bolton Abbey, will serve for an example), the hard grit rocks rise in picturesque and sometimes wild-looking escarpments. If we ascend these from the valleys below we shall find the upper part comparatively level or undulating, and stretching away for miles in every direction, until its western and northern continuity is interrupted by the occurrence of another valley which has been eaten out like that first mentioned. These fragmentary plateaux are the moorlands proper, the home of the grouse. Each table-land will be known by a different name, generally one of locality. But all these western and northern Yorkshire and Lancashire moors are as like each other in their physical scenery as the patterns of the same carpet. We have the same billowy expanse of whin, bilberry and heather bushes, diversified by the rough-and-tumble distribution of rock fragments and boulders, looking as if they had been sown broadcast by some giant. In the later summer-time the surfaces of these moors are a floral glory,—a mass of diversified colour. In the winter they are dark and gloomy, and in keeping with the mournful wintry sky. It is in their latter aspect that Charlotte Bronte best loved to describe the Yorkshire moors in the neighbourhood of Bradford. Wordsworth's "Bardston Tower" stands on the solitary plateau of Rumbold's Moor, on the escarpment above the beautiful Wharfedale.

Not unfrequently the surfaces of our moorlands have been softened by the long-continued growth and accumulation of vegetable matter, which has at length formed an impure peat of many feet in thickness. Sometimes we are surprised at the number of tree-trunks and stools which occur in this peat, although at the present time none but dwarf alder, birch, or juniper will grow on the moorlands. A little investigation shows us that these peat-buried trees are none of them of large size, and that they are also dwarfed. But their number is great, and shows us that the bare surfaces of these rocks were once truly forest-clad. In East Lancashire we have an undulating track of uncultivated moorland, cultivated pasture, and valley, and the entire country goes by the name of Rossendale *Forest*, although there is now hardly a tree to be seen except in the valleys. But the history of the district shows us that as late as the time of the Tudors, the moorland slopes were actually well wooded, and the peaty soils are stored with the remnants of some of the ancient ancestors of this forest.

In places these peat beds have been recently subject to wear-and-tear by surface streams and floods, and these have ploughed out deep trenches and miniature black gullies, in whose sides we see the gnarled roots, trunks, and branches of trees imbedded. As the original surface of the rock is also thickly covered with rocks and boulders, it follows that these surface streams have to wind in and out among them, and in this way they have cut a series of the most baffling and (to the sportsman) annoying channels, for they have to be leaped over or waded every few yards. These streams eventually combine, to form a tributary of importance to the stream which we shall surely find in the nearest valley, and whose dark, peat-stained waters will show that they have drained from the moorlands and been dissolving away the peaty soils. Who that has wandered over Rannock Moor, the famous "sketching country" of Landseer, has not been struck with the manner in which the thick layer of peat has been thus cut up into boulder-strewn patches, separated from each other by dark streams? Here and there, in the lowest lying parts to which the drainage converges, there is a series of small peat-embosomed tarns, lying together like beads on a string. But everywhere on the outside edge of the great plateau occupied by Rannock Moor, the larger streams drain off into the lochs below.

To the east of York we find another set of escarpments rising above the extensive plains. On ascending them we discover another series of plateaux, forming the Eastern Yorkshire moors. But we at once perceive a difference, both in the physical geography and in the distribution of plants. These eastern moorlands lie on Oolitic limestones, and they are therefore richer in grassy pastures. Here and there we find widened fissures and "pot-holes," whilst the ice-sown "boulders" to some extent assimilate the surface appearance to what we have been accustomed on the Millstone-Grit. Angular fragments, however, are much scarcer, and altogether absent in places. Nor is there anything like the wild expanse of heath and whin which gives the western and northern moorlands more than half their charm, whilst the greater abundance of cattle further helps to

tone down the solitude and make us feel that we are
in the neighbourhood of home.

In the Peak district of Derbyshire the hills are
formed of Millstone-Grit and Carboniferous Limestone, the intervening thick bed of soft black Yoredale
shales having been eaten away to form the verdant
valleys which now separate the grit hills from the
limestone. If we follow up some of these valleys to
their head we shall find they are *culs de sac*, "blind
alleys," which have been unquestionably denuded
out of the soft shales, for the "head" of the valley
is an outcrop of these soft shales, and very likely an
escarpment, weathering at the present time so fast
that it cannot be greened over either by moss or
lichen, but is always bare. Such a "head" of a
valley is the celebrated "Mam Tor," near Castleton,
in Derbyshire,—a hill of Yoredale shales, weathering
away so fast that for generations it has been popularly
known by the name of the "Shivering Mountain."
The tradition runs that it is "always wasting away,
but never getting less!" The fact being that the
quantity of waste or *débris* accumulating at the foot is
evident to the eye, but the decreased size of the hill
from the operation is not so apparent, although we
know it is equally certain.

The Millstone-Grit hills and the Limestone hills of
Derbyshire often occur on either side of such valleys,
and the tourist cannot but be struck with the difference of the vegetation covering the surfaces of each.
On the former we have the chaotic wilderness of
rough crags, rocks, and boulders set in a framework
of heather and gorse ; on the latter, according to the

angle of the dip of the strata, we have terraced out-crops of limestone. The atmosphere acts *chemically* as well as mechanically upon limestone, for the carbonic acid of the rain-fall dissolves away the Limestone surface on which it falls. Let any one examine the almost-naked surface of the Carboniferous Limestone of the barony of the Burren, in county Clare, and he will be surprised at the manner with which weather action can affect the scenery of a district. In the Burren district the annual rainfall amounts to about 54 inches; but it usually descends in such fine showers that very minute drops must tell upon the limestone, and thus its surface is being so constantly weathered that vegetation has no opportunity of growing. The numerous fissures in the limestone, however, are constantly receiving a " wash" from the surface; and in them we find veritable rock-gardens of ferns and flowering-plants and grasses, some of them not to be found elsewhere in the British islands. The dissolving action of the weather is little, or not at all felt upon sandstone rocks; hence their surfaces remain rough, and lack that general smoothness which on limestone rocks has been chiefly produced by the dissolving action of the rainfall. On the limestone, also, we get a short, sweet greensward, of which the mountain sheep are especially fond. A soft, spring turf everywhere underlies our feet, except where some harder mass of rock crops out above its surface.

All limestones are liable to joints and fissures, due originally to the contraction of the strata when they passed into the solid state. The harder the limestone rocks, and, as a rule, the more numerous and deeper are these joints. Very frequently such joints serve as lines of superficial drainage. All the waterfall trickles down their sides, and thus they are also liable to chemical as well as mechanical denudation. In this way some of the limestone walls have been gradually weathered asunder for ages, until they are now many feet or even yards apart; and thus we have those remarkable passes formed which occur only in limestone districts (and particularly in the Carboniferous formation), like the winding gorge of the Cheddar Cliffs, the "Winnats" near Castleton, and many others. It is probable that the "Dales" of Derbyshire, or rather of the limestone area of that county, such as Miller's Dale, Monsal Dale, Dovedale, and others, were formed in this manner.

The surface of this limestone is liable to pits and holes, caused by the chemical and mechanical wear-and-tear of the solid rocks by the water which has flooded after rainfalls, descending some chink or fissure, and wearing it irregularly wider. Of such origin are the famous "Eldon Hole" in the Peak of Derbyshire, concerning which local tradition has had so many strange stories to tell; the "swallow holes," and "pot-holes," found on the surfaces of the limestone in Lancashire and Yorkshire, have originated in a similar manner. In Derbyshire a railway station on the Midland is called "Dove-holes," —the name euphemistically expressing the number of swallow-holes which occur in the district.

The dissolving and excavating work of water is most completed when it has disappeared down these crevices. It finds its way under ground, wearing and dissolving the rock-walls as it passes along, and in this way forming subterranean passages or caverns which often extend for miles. The caverns so abundant in limestone regions, but which occur high up along the hill-sides, represent some ancient level of drainage, that has been altered by upheaval, or by the deeper excavation of the main valleys below. Sometimes these underground passages have been excavated whilst surface wear-and-tear has been going on, until at length only a thin roof has shut in the line of caverns below. Let that wear down or fall in, and the underground caverns are then excavated into dark and gloomy winding gorges, into which, however, daylight for the first time finds its way. After this fashion was formed the narrow, upward-winding, and picturesque gorge at Castleton, known as " Cave Dale ;" and, perhaps, many another of the narrower and abundant Derbyshire " dales."

On the boundaries of the soft Yoredale Shales, and the Carboniferous Limestone near Lisdoonvarna, co. Clare, we can see that the upper roofing of the limestone caverns is very thin. Here and there we may observe a portion actually broken in, down which the rains wash with great force during a shower. After a few days' heavy rains the caverns become flooded with water, part of which wells up through these roof-chinks, and fills up the depressions adjacent, converting them into temporary tarns. It may be that many of the narrow, craggy, and precipitous ravines and gorges which are so characteristic of the scenery of the Carboniferous Limestone in hilly regions are

nothing but ancient caverns whose roofs have been
worn off!

Even streams of considerable size will sometimes
disappear through "swallow-holes," and be lost to
sight at the surface. They will then wander underground for miles, receiving subterranean tributaries,
and eventually emerging as larger rivers. Of such is
the stream which may be seen issuing from the mouth
of the Peak cavern. In the Yorkshire limestone, the
streams which emerge from caverns are known as
"Caldes."

In their recently published work on the botany,
physical geography, and geology of "West Yorkshire," Messrs. Lees & Davis speak as follows of
these and associated natural phenomena :—"Amongst
the characteristic features of the scar limestone are
the great fissures and cracks, which were probably
formed during the consolidation of the rocks, and
which often indicate natural joints of considerable
extent. Many of these subterranean passages are
occupied by torrents of water, which, being collected
on the hill sides, disappear in openings in the limestone plateau, often called 'pot-holes,' and after
pursuing a devious course, in some instances two or
three miles in length, reappear at a lower level, or at
the base of the cliff, in copious streams. Numerous
instances of such phenomena might be cited as occurring on the slopes of Ingleborough, Penyghent,
and in other localities. The stream which emerges
in Clapdale is swallowed four hundred feet higher
up by Gaping Ghyll, a terrific chasm on the limestone plateau of Ingleborough. The river Aire, in

its passage from Malham Tarn, sinks into a cleft in the limestone, and emerges at the foot of the precipitous Cove ; and the river Nidd follows a subterraneous course above Pateley Bridge. The ebbing and flowing well above Giggleswick Scar is an instance of a different character."

In the limestone tract of the Peak of Derbyshire, the "ebbing-and-flowing" wells are not uncommon ; indeed, one little town among the hills, Tideswell, owes its name to such a phenomenon, which, however, has ceased to operate within the last few years. Speaking further of the chemically-dissolved holes, fissures, &c., which diversify the Yorkshire limestone moors, Messrs. Lees & Davis go on to say: "Numbers of the pot-holes do not form channels for water, but have probably fallen in over chasms existing below the surface. They are often in lines, of great regularity, in the form of inverted cones, their sides grown over with grass, sometimes with an open chasm of immense depth at their apex (as at Alumpot, above the village of Selside), but in the majority of cases the shelving sides converge to a point without exhibiting an orifice. Many caves and caverns also exist in the limestone. Some are partly open to the surface, as Weathercote Cave, in which is a magnificent fall of water 75 feet high ; and the Victoria Cave, at Settle. Others are only to be seen by passing a narrow horizontal or vertical entrance. Ingleborough Cave, a short distance above the village of Clapham, is one of the finest and most remarkable in the district. It was formerly the watercourse of the stream from Gaping Ghyll hole, its narrow passages and large chambers being often nearly filled up with pebbly and sandy sediment. It has been explored to the distance of nearly half a mile from the entrance. At the further extremity the rush of waters can be heard, which, however, having been diverted into another channel, make their exit at a lower level, a few yards to the right of the opening of the cave. The sides and roof of the cave are full of minor fissures."

In the Burren, west of Ireland, remarkable sub-terranean rivers are very common, especially in the neighbourhood of Gort.

Mr. G. H. Kinahan has described the underground rivers of the mountain regions of Ireland in his geology of that country, and his description of existing physical operations throws great light on the origin of mountain gorges, ravines, &c. He says: "In some of the mountainous districts, such as those of Cork, Kerry, Galway, and Clare, streams are for short distances subterraneous, or lakes have subterranean outlets. In Slieve Aughter, co. Clare, there are subterranean passages that have been dissolved out along the joints in the peculiar limestones which in places form the basal bed of the Old Red Sandstone. In the upper limestones, however, are found the great system of subterranean rivers, with which are associated lakes, turloughs or blind lakes, and holes, locally known as ' sluggas,' ' swallow-holes,' or ' pot-holes.'

" The *turloughs* and *sluggas* are remarkable features. The latter are due to portions of the rock falling in to the underlying vacancies made by the subterranean rivers ; sometimes the courses of the rivers can be

traced by them ; in some localities they are dotted
irregularly over a tract of ground as if the ground
were traversed by numerous subterranean streams.
Some are abrupt deep holes, others open into shallow
hollows ; and when the water during floods rises in
the latter, it overflows the adjoining land, forming the
turloughs, which are usually lakes in winter, and
callows in summer. The subterranean channels seem
to have commenced by the water dissolving out vacancies along joints and other weak portions of the rocks,
more in some places than in others ; as the channels
and streams increase in size, vents (*sluggas* or pot-
holes) break in from the surface, and stones and such
like force their way into the stream, thus causing
the currents to act more like surface ones, abrading
as well as dissolving away the rocks. A *slugga* is
usually shaped like an hour-glass, although some have
perpendicular sides ; they seem always to be formed
from below. The water of the underground rivers
and streams during floods acts upwards, in the joints
and other weak portions of the roof, till eventually
openings are formed connected with the surface,
while afterwards the holes are more or less increased
in size by meteoric abrasion."

From this matter-of-fact description, the geological
student can easily perceive that nothing is wanted but
TIME to enable such natural operations to eat away
the winding gorges and " dales," which are such a
characteristic feature in the scenery of most limestone
regions.

Mr. Kinahan goes on to describe these interesting
phenomena, and as we read we seem as if we were

witnessing the important act of gorge-making. He
says: "In the limestone country between the estuary
of the Shannon and Killala Bay in Sligo, the rivers
and streams are more or less subterranean, the best
developed system being in the neighbourhood of
Gort. Lough Cooter, near Gort, receives the drainage
from most of the north-western portion of Slieve
Aughter. The Beagh river flows out of Lough Cooter,
having for about two miles an open course, when it
disappears in the limestone under a drift cliff. But
its course can be traced by 'sluggas,' called the
Devil's Punch Bowl, the *Black Weir*, the *Ladle*, and
the *Churn*, to Pollduagh, a cave out of which it
again comes to the surface. After this it is an open
river for about three miles, when it sinks again north-
east of Killarten, but rises west of that village to sink
and rise *several times*, till eventually it finds its way
into Coole Lough. From Coole Lough the water
finds its way by subterranean passages to the sea.
Part seems to flow into Galway Bay at Kinvarra, a
distance of six miles as the crow flies; but the water
does not all come out here; some may have vents in
Galway Bay, but it is popularly believed that a considerable portion of it goes south into the lakes that
join with the Fergus; thus flowing into the estuary of
the Shannon.

"In connection with this river system are Caher-
glassaun Lake, the small turloughs in its vicinity,
and a small lake west of Gort, which rises and
falls with the tide in Galway Bay. This rise and
fall cannot be observed during floods, on account
of the expanse of water, when it is so small as

to be scarcely perceptible ; but at other times it is
most marked.

"To the south of this area is the water basin of
the Fergus. The main stream bursts as a fully-
formed river out of a limestone cliff under the coal
measures, a few miles north-west of Ennis. To the
east of the basin of the Fergus, at Killannon, are the
celebrated *Tomines*, or immense natural vaulted passages in the limestone, through which the river
Ardsolla winds an extraordinary course. These present a scene of magnificence never to be forgotten
by those who have viewed it."

So that whilst we are indebted to the meteorological
agents of former periods for many of the physical
features of modern landscapes, it will be seen that
we have equally active agencies now in operation,
whose influence will perhaps be experienced in
periods yet to come !

CHAPTER IV.

OUR ALPINE FLOWERING-PLANTS.

The Great Ice Age, or "Glacial Period," and the Plants which
then prevailed in Britain—Arctic Plants on British Mountains
—On the Alps, Pyrenees, &c.—Darwin on the Origin of
our British Alpine Flora—Arctic Plants only Floral Refugees
—The Mountain Avens—Alpine Saxifrages—Snowdon Pink
—Alpine Lychnis, Cerastium, &c.—Arenaria Norvegica—
Alpine Saginas — Mossy Cyphel—Alpine Hawk-weeds—
Saussurea—Mountain Everlasting—Mountain Cud-weed—
Mountain Meadow-rue—Nuphar Pumila—Water Awl-wort
—Lobelia—Alpine Willow-herbs — Bird's-eye Primrose —
Mountain Sorrel—Alpine Lady's Mantle—Draba, Arabis,
Teesdalia, &c.—Delights of Alpine botanizing.

WE have already alluded to the fact that amid the
most lovely and inaccessible crags of our highes
mountains, there bloom the rarest of our British
flowering-plants.　Some of these rare plants are worth
gathering, if only for the sake of their floral loveliness, but particularly so for other reasons of higher
intellectual importance.　For these rare Alpine
flowers are indicators of a former condition of physical geography, without whose influence they could
never have found their way to their present habitats.

In a previous chapter we have given a rough outline
of the physical conditions which prevailed in Great
Britain during the Glacial period.　We saw how our
mountains and hills abound in proofs, both of a for-

mer continental and a subsequent glacier condition of ice. Even at the close of this long-continued "Great Ice Age" the cold was exceedingly great. Our islands at that time were connected with the continent of Europe, they had not been isolated by the submergence of the low-lying plains. When the ice cleared away from the surface, wherever plants could grow we may be sure they were only those which could endure the cold. Indeed, the extension of Arctic conditions of climate during the Glacial period brought with it an extension of the Arctic flora. This flora not only extended over the British area, but even much farther to the south of Europe. When the warmer climatal changes ensued, that gradually brought about the present state of things as regards the weather, the difference rendered the cold-loving or Arctic flora unable to compete with the warmth-loving plants which made their way into Great Britain from the South-east, across the plains of what are now the bottoms of the German Ocean and the Irish Sea. Accordingly, the Arctic flowers ceded the lower grounds, the only places remaining open to them for shelter being the cold flanks and summits of the higher mountains, where they would not be likely to be expelled by the newly-introduced low-land and warmth-loving plants. Hence it is that we find the remnants of this ancient flora solitarily blooming in our mountain fastnesses, lonely, but eloquent testimonies to the former severe conditions which brought them over the areas where they still linger.

We find these Arctic plants not only growing on the sides and summits of our English, Welsh, Scotch,

and Irish mountains, but on the margins of continental glaciers as well. On the Faulhorn, in the
canton of Berne, at 9,000 feet above the sea-
level, there grow one hundred and thirty-two species
of flowering plants of which fifty-one are common to
Lapland and eleven to Spitzbergen. In the Enga-
dine, a high valley in the canton des Grisons, there
are found eighty species of plants unknown to the
rest of Switzerland, but which are very common
within the Arctic circle. Taking the *Alpine* flora of
Switzerland as a whole, we discover that out of a
total number of three hundred and sixty species, one
hundred and fifty-eight are common to Scandinavia
and northern Europe generally. The relation of
the European Alpine flora to that of the Arctic
regions may also be seen by reversing this comparison. Thus, out of six hundred and eighty-five species of flowering plants found in Lapland, one hundred and eight are also known as growing on the
Swiss Alps. The extension of the Arctic flora
southerly, during the Glacial period, obtained as far as
the Pyrenees, where we meet with sixty-eight species
of plants common to Scandinavia.

Darwin describes the origin of our Alpine flora in a
similar way. Speaking of the close of the "Great
Ice Age," he says :—" As the warmth returned, the
Arctic forms would retreat northwards, closely followed up in their retreat by the productions of the more
temperate regions, and as the snow melted from the
bases of the mountains the Arctic forms would seize
on the cleared and thawed ground, always ascending
as the warmth increased, and the snow still further

disappeared, higher and higher, whilst their brethren were pursuing their northern journey. Hence, when the warmth had fully returned, the same species, which had lately lived together on the European and North American low-lands, would again be found in the Arctic regions of the Old and New Worlds, and on many isolated mountain-summits far distant from each other."

"Thus we can understand the identity of many plants at points so immensely remote as the mountains of the United States and those of Europe. We can thus also understand the fact that the Alpine plants of each moutain-range are more especially related to the Arctic forms living due north or nearly due north of them ; for the first migration when the cold came on, and the re-migration on the returning warmth, would generally have been due south and north. The Alpine plants, for example, of Scotland, as remarked by Mr. H. C. Watson, and those of the Pyrenees, as remarked by Ramond, are more especially allied to the plants of Northern Scandinavia ; those of the United States to Labrador ; and those of the mountains of Siberia to the arctic regions of that country." In like manner it will be found that the Alpine plants of all our highest English hills and mountains are related to those of Scotland and Scandinavia ; or rather (and here we also include the Alpine flora of Wales and Ireland) they have the same common origin.

Can the reader now marvel that so many of our mountain flowers are clothed with new wonder for a thoughtful botanist ? They have found a refuge from

the low-land climatal invasions in those very highlands where the aboriginal Britons also sought shelter
from the overwhelming encroachments of Roman,
Saxon, Dane, and Norman ; and the old British
speech is heard as
Cymric, Gaelic, or Erse,
chiefly in the very moun-
tains where floral refu-
gees also sought and ob-
tained an uncontested
freedom.

Let us now proceed
to note the more re-
markable of these Alpine
plants. Although some
of them are exquisitely
beautiful, others are
very simple and unpre-
tending.

But even this scanty,
historical Alpine flora
has its likes and dislikes.
Where the mountains
are formed of Carbonif-
erous limestone we may
expect to find the white
Mountain Avens (*Dryas
octopetala*), whose woody

MOUNTAIN AVENS (*Dryas
octopetala*).

stem and pretty crenated evergreen leaves (resembling
those of the oak, whence its generic name), often
completely mat the ground where it takes up its
abode. Nowhere have we seen it growing in such

luxuriance as on the limestone hills which terminate in the bold cliff of Black Head, in the west of Ireland. It is also found in Wales, Scotland (where it ascends to a height of 2,700 feet), and the tops and flanks of the limestone hills in the West Riding of Yorkshire. Professor Kerner has recently pointed out the purpose of leaves of the character possessed by the Mountain Avens. He says: "In many plants the foliage is of thick and leathery consistence, and this acts as a security against injury from cattle. The wide tracts in the Alps which are seen covered with evergreen carpets and shrubby thickets of *Azalea procumbens*, *Dryas octopetala*, *Empetrum*, &c., and other characteristic plants, are avoided by sheep, and also by chamois. It is exceptional to find the leaves of these plants mangled by grazing animals, and we never find them completely destroyed." The rambler cannot fail to recognise this flower from its eight snow-white petals. Chief among these cold-loving plants is represented the genus *Saxifraga*, literally "stone-breakers," from the old belief (based on the rocky habitats of so many species) that the roots of some of these plants could penetrate the hardest rocks. At least *seven* species of this plant are characteristically Alpine or sub-Alpine. We never find them except on elevated regions, and there under slightly different conditions. Thus *Saxifraga stellaris* loves to grow beside the cool mossy mountain-springs and rills, where its starry white flowers can be easily recognised from their resemblance to that rare but indigenous plant which English cottagers delight to use for the borderings of their gardens,

the elegant "London Pride" (*Saxifraga umbrosa*).
This latter plant was thought to be indigenous only
on the mountains of south-western Ireland, where it
is known among the natives by the name of
"St. Patrick's Cabbage." Recently, however, it has
been found growing on the hills about Settle, in the
West Riding, and there cannot be the slightest
doubt it is truly indigenous to this new locality.
Another of the commoner mountain Saxifrages bears
bright yellow flowers, whose petals are richly adorned
with orange-coloured spots, on downy and glutinous
stalks (*Saxifraga aizoides*). This species ought to be
looked for in the stony places beside the Alpine rills.
Both it and *S. stellaris* are exceedingly abundant
along the wayside as we cross the top of Kirkstone
Pass, to descend into the valley of Brotherswater.
The *stellaris* is abundant beside rills everywhere in
North Wales at a certain height, but the Yellow
Saxifrage has not as yet been found in that country.
The latter is extraordinarily abundant in the High-
lands, sometimes covering large spaces with its crowded
linear leaves.

Rarer species of Alpine Saxifrages are limited to
certain mountains, as the Purple Saxifrage (*S. oppo-
sitifolia*), whose solitary, bright purple flowers love
to grow in dark places on the crags of Snowdon, and
on the tops of the highest cliffs of the Highland
mountains, to an altitude of 4,000 feet. Not far from it
(for this plant loves the same inclement exposure)
may be found *Saxifraga nivalis*, easily recognised,
not only by its many white blossoms, but also from
its leafless stem rising from the leathery leaves. It

grows on the summits of the higher Welsh and
Scotch mountains, in places so abundantly as to
cover the ground. Two other species of Saxifrage
(*S. cernua* and *S. cæspitosa*) are found, the former
only on the schistose rocks near the top of Ben
Lawers, at a height of 4,000 feet ; and the latter in
scarce patches on Ben Nevis and Ben Avon, between
3,000 and 4,000 feet above the sea-level. *Saxifraga
hypnoides*, although it grows to as great a height on
our mountains as 4,000 feet, can nevertheless exist
at a much lower altitude, for it abounds in the moist
places on the limestone rocks of Derbyshire, and
extends as far south as North Somersetshire.

The natural order of plants to which our pinks,
chickweeds, and stitchworts belong, is also remarkable for the number of Alpine species it includes.
Of these the most noteworthy is the Moss Campion,
sometimes called the "Snowdon Pink" (*Silene
acaulis*). The flower is not more than four inches
in height when it expands its lovely purple or white
flowers (for we find it of both colours). It grows amidst
dense tufts of small linear leaves, which, on some of
the higher Scotch mountains, as for instance Ben
Ledi, completely carpet the ground where it is sufficiently moist. The Alpine Lychnis (*Lychnis alpina*)
grows less abundantly, but in similar places. Its
rosy flowers adorn the summits of Hobcartin Fell, in
Cumberland ; and Little Kilrannock, in Perthshire.
The other Alpine plants of this order are more nearly
related to the chickweeds, such as *Cerastium alpinum*,
found at great heights on the mountains of Wales,
Westmoreland, Cumberland, and Scotland. A much

rarer plant is *Arenaria rubella*, found on the rocky summits of Ben Hope and the Breadalbane mountains in Scotland, but not at a less height than 2,500 feet. *Arenaria Norvegica* is, as its botanical specific name implies, another of these interesting arctic plants which physical changes have forced to become *Alpine.* It is almost limited in its range, however, to the high grounds of Shetland and Orkney. *Sagina saxatilis* and *S. nivalis* are two species of allied

THE MOSSY CYPHEL (*Cherleria sedoides*).

plants, also bearing inconspicuous flowers, which are to be met with on Scotch mountains; the latter being most abundant on Ben Lawers, and on the hill-tops in Skye. The Mossy Cyphel (*Cherleria sedoides*, or *Arenaria cherleria*), as it is now called, grows only on the lofty Highland mountains, where its leaves form dense, mossy-looking green cushions, out of which the green sessile flowers peep forth.

The *Hawk-weeds* (as the composite, dandelion-like flowers are called, from the old belief that hawks in some mysterious manner derived their keen-sightedness from the virtues of these plants) yield us a good many characteristic Alpine species. Although not exactly one of them, still the Purple Sow-thistle (*Mulgedium alpinum*) is not far removed. It is a very beautiful and very rare plant, flowering in the late summer on the wet Alpine rocks of Forfar and Aberdeen, and perhaps most abundant on the crags of Lochnagar, but never met with at a less altitude than 2,000 feet above the sea-level. *Saussurea alpina* is a commoner plant, bearing large dense clusters of purple flowers with blue anthers, which look like thistle-plumes set in a large flower-cup. The leaves are slender, and are woolly on their under surfaces—a very common occurrence in mountain plants. The *Saussurea* grows on the high mountains of North Wales and the Lake district, as well as on those of the Highlands, ascending to nearly 4,000 feet. The Mountain Everlasting (*Antennaria dioica*) is also a composite plant peculiar to high regions, loving most to grow on the dry mountain heaths, where it is sometimes very abundant. It is seldom met with, however, at greater altitudes than 2,000 feet. It is peculiarly abundant in Scotland, but less so, even on the hills, nearer the south. Nearly allied to this plant is *Gnaphalium alpinum*, or Mountain Cud-weed, which loves similar habitats.

Of the Alpine Hawk-weeds about fifteen species are peculiar to the Scottish mountains. The Alpine Hawk-weed (*Hieracium alpinus*) grows on the mountains of North Wales, the Lake district, and Scotland. *H. nigrescens* ascends in Scotland to upwards

of 4,000 feet. The copses and damp shady places
along the mountain and hill sides, especially in Scotland and the Lake district, are usually peopled by one
or more species of these northern hawk-weeds. No
fewer than six species grow on the hills of the West

THE MOUNTAIN EVERLASTING (*Antennaria dioica*).

Riding of Yorkshire, one species (*H. Gibsoni*) being
apparently limited to the limestone hills near Settle.

The Mountain Meadow Rue (*Thalictrum alpinum*)
is a well-known and tolerably wide-spread mountain
plant, generally to be met with in the boggy places at
altitudes up to 4,000 feet. Its bright green, rue-like
leaves, and slender stems drooping gracefully with the
weight of their scant cluster of palish yellow flowers,

give it a most elegant appearance. Perhaps this very plant grows alongside the silent mountain tarns, where, in Scotland, there may be found one of the rarest and loveliest of our British flora, the small yellow Water-lily (*Nuphar pumila*). If such a lake or tarn be not too elevated, say not more than a little over 2,000 feet above the sea-level, we may possibly find the Water Awl-wort (*Subularia aquatica*), a little cruciferous plant, growing where the bottom of the lake is of a gravelly nature. It is met with in the mountain tarns both of the Lake district and of Scotland. In the same habitats as the Awl-wort (although as a rule preferring lake-beds at a less elevation) there may be obtained *Lobelia Dortmanna*. It is found in these situations in the tarns of the Shropshire hills and of the Welsh mountains, but it is particularly abundant in those of the Lake district, where it descends even to the levels of such large lakes as Windermere, and grows also along the gravelly bottoms of such shallow streams as the Brathay and Rothay, which empty themselves into the lake. Hardly a mountain tarn is without this most beautiful aquatic plant, with its scape of drooping pale liliac flowers rising sometimes a foot or more above the surface of the water.

In the neighbourhood of mountain rills and very moist places the pretty little Alpine Willow-herb (*Epilobium alpinum*) may be found, and easily identified by its two pubescent leaves and bright rose-purple flowers. It grows over most of the hills and mountains north of Durham, and into Scotland, ascending as high as 4,000 feet. *Epilobium montanum*, another hill plant, never ascends to 2,000 feet, and its flowers are of a pale purple. The Mountain Sorrel (*Oxyria uniformis*) is an abundant mountain plant, especially in Scotland, where it may be found growing alongside every rill and in most damp places, and easily recognised by the fleshy, kidney-shaped leaves, which may be further proved by their acid taste. The Bird's-eye Primrose (*Primula farinosa*) also loves damp and moist habitats, where its presence is soon made known to the keen nostril of the botanist by its delicate musk-like smell. It is a lovely plant, although the flowers are small, for the latter are of a pale purple tint with a yellow centre. The under surfaces of the leaves, which are arranged in a rosette-shape, are covered with a fine brimstone-coloured meal or dust, whence the specific name. This plant is especially abundant in the Lake district, and nowhere is it more plentiful than in the lovely valley leading upwards from Rydal Water to the top of Fairfield—a route well worth trying, if only for the sake of the many rare plants to be met with, and the splendid examples of ancient *roches moutonnées* which are to be seen. Another most abundant mountain plant in Cumberland and Scotland is the Alpine Ladies' Mantle (*Alchemilla alpina*), whose elegantly-cut leaves can be easily identified from the allied species so abundant in the hilly pastures of the northern counties (*Alchemilla vulgaris*) by the silvery look produced by the downy hairs on the under surfaces of the leaves. It grows wherever the soil is moist enough, descending to the very bases of the mountains.

In the dryer and rockier of Alpine places we may find a few diminutive and unattractive cruciferous plants, which are chiefly remarkable to botanists because they are not found growing elsewhere than on our high hills and mountains. Such are *Draba rupestris*, seldom found at a less altitude than 3,000 feet ; so that it is necessarily rare and local, having been only met with on Ben Lawers, the Cairngorm mountains, and Ben Hope, in Scotland, and Ben-bulben, in Ireland. *Draba incana* is a commoner Alpine species, found on most of the high mountains of Wales, Cumberland, Scotland, and south-western Ireland. *Arabis petræa* is another mountainous cruciferous plant found growing at high localities on the Welsh mountains, and on those of Skye and Shetland ; as well as on Braemar and the Cairngorm mountains. The naked-stalked Teesdalia (*T. nudicaulis*) is an Alpine, cruciferous plant, abundant in Ennerdale, in the Lake district, and elsewhere.

NAKED-STALKED TEESDALIA
(*Teesdalia nudicaulis*).

The pursuit of these botanical rarities is fraught with the highest enjoyment. Whilst hunting for them in nooks and crevices among the crags, the adventurer perhaps startles many an equally rare mountain bird or mammal. He sees many a weathered fossil standing forth amid the rocky ribs of the hill-sides, or notes peculiar mineral veins and other details of mineralogical structure. Grottos, worthy of the

fabled fairies which popular belief still associates with
the mountains, reveal their treasures of rare mosses,
ferns, and lichens only to his privileged eyes.   The
mists which gather around the peaks above him add
to the sublimity and loveliness of the scene.   Below
him the billowing hills roll away to the farthest horizon, and their smiling valleys glitter in the sunshine
which bathes them, whilst the embosomed lakes gleam
like sheets of molten gold ! It is well for even the best
and strongest of us to feel what Wordsworth termed
"the power of hills," at least once in our lives.
Summer rambles among the mountains, perhaps more
than any other form of outdoor enjoyment, lead the
heart to meditate upon the Power whose majesty
seems to find the fittest footstool thereon.

CHAPTER V.

OUR ALPINE FLOWERING PLANTS.

(*Continued*).

Alpine Forget-me-not—Mountain Gentians—Mountain **Speed-wells** — Baldmoney—Alpine Milk-vetches — Oxytropis —Lathyrus—Sibbaldia—Scottish Asphodel—Bog Orchis—Whortleberry, Cowberry, **Cranberry, Bearberry** — Andromeda—Dabeocia — Azalea — Winter-Greens — Epimedium—Linnæa —Trientalis—Cloudberry, Crowberry, &c. — Rose-root—Alpine Butterwort— Alpine Bistort — Alpine Scullcap — Alpine Fleabane—Alpine Enchanter's Nightshade—Alpine Willows—Dwarf Birch—Juniper—Bog Myrtle—Scotch Fir—Its Geological Antiquity.

ONE of the Alpine floral gems, to obtain which many a botanist has perilled his limbs in climbing the almost inaccessible crags amid whose moist places it loves to nestle, is the Alpine Forget-me-not (*Myosotis alpestris*). It has been scantily met with in the mountains of Westmoreland, and at between 2,000 and 3,000 feet up Teesdale. Ben Lawers, however, is its chief British station, and there we find it growing 4,000 feet above the sea level. Two species of Mountain Gentian (*Gentiana nivalis* and *G. verna*) are also regarded as good "finds," especially the latter, which loves to grow most on wet limestone rocks at sub-Alpine altitudes, and is found at between

2,000 and 3,000 feet in Upper Teesdale, and still more abundantly on the hills of Mayo and the Carboniferous limestone of County Clare. Its bright blue corolla is an inch in diameter, and therefore a very conspicuous object. *Gentiana nivalis* has its smaller but pretty bell cut into five segments. It is found on the heights of the Breadalbane and Clare mountains, at close upon 3,000 feet. Both these species of Gentian are remarkable for the manner with which the stigma or top of the pistil is flattened out, so as to cover in the tube of the flower like a lid. Kerner has shown that this device is to prevent small and to the plant *useless* insects from getting at the nectar. The purpose

MOUNTAIN FORGET-ME-NOT
(*Myosotis alpestris*).

of the latter is to reward those larger insects which are beneficial to the flowers by crossing them. An insect wishing to suck the nectar at the bottom of the tubes of these Gentians must insert its proboscis by the margin of the stigma; and to do this it must push away the folded corolla which enfolds it. This, however, is a task which only the largest and most useful insects can achieve, and thus the sugary reward is to the strong.

Two other Alpine flowers worth noting are Speed-wells (*Veronica alpina* and *V. saxatilis*). The former is an erect little plant, with ovate leaves, and it bears a few pale blue flowers, which must be sought for beside the mountain rills at not less than 2,000 feet above the level of the sea. It is a very rare species, however, and entirely limited to the Scottish Alps. *Veronica saxatilis* is a much prettier plant, for its flowers are of a most attractive bright blue colour. Like its relative, it is also very rare, and occurs only on or near the summits of Breadalbane, Ben More, and Ben Cruachan.

The drier mountain pastures have also their characteristic Alpine plants. One of the most remarkable of these is the "Baldmoney," as it is called by the Highlanders (*Meum athamanticum*). It is also known as "Spignel" and "Mew"—its several names being a sure sign that it is not uncommon. The flower-stalk is somewhat tall and ragged-looking, and is covered with numerous nuclei of yellowish tinted flowers, for it is a member of the *Umbelliferæ*. Like many other of this natural order, the plant is aromatic, and the carrot-shaped root was formerly chewed as a carminative. We find this plant growing in alpine pastures all the way northwards from Wales and Lancashire as far as Moray. It has not been found in Ireland.

In the drier altitudes of the Scottish mountains, especially in Aberdeen and Forfar, and on Craig-an-dal, near Braemar, at the head of Gendole, may be

BALDMONEY (*Meum athamanticum*).

found the rare and beautiful Alpine Milk-vetch (*Astragalus alpinus*). Its drooping flowers are of a pale blue, tipped with purple. *Oxytropis* is a genus of similar leguminous plants, whose two British species both affect mountain-heights. The rarer is *Oxytropis campestris*, which bears pale yellow flowers tinged with purple, and hairy pods. The other species is *Oxytropis uralensis*, and this we find growing in mountain pastures almost everywhere in Scotland, at heights below 2,000 feet. Its flower-heads bear from six to ten pale purple flowers, clustered together, each flower about three-quarters of an inch long. The root-stock is stout and strong, and covered with woolly hairs. *Lathyrus niger* is another Alpine member of the same order, confined to the Scottish mountain valleys. The tourist may gather it abundantly in the well-known Pass of Killiecrankie, and at Craiganain, near Moy House. It takes its specific name from its habit of turning black when dry. The flowers are of a livid purple colour, gradually changing to blue. The stems are tolerably erect, and from one to two feet in height.

On the Scotch mountain-summits we may possibly find the very rare Spider-worts, but at any rate those are the places to look for *Sibbaldia procumbens*, whose smaller yellow flowers are apt to be overlooked or mistaken for those of its near ally the common Tormentil. It usually affects stormy habitats, at heights of from 1,500 up to 4,000 feet, and its distribution over them is confined to the hills and mountains from Peebles northwards.

One of the charms of botanizing along a mountain-

range is the diversity of physical conditions which
the rambler comes across. Here is a patch of oozy,
treacherous, and forbidding ground, telling plainly of
some spring which wells up and produces a local
quagmire. But the place has its own suite of plants,
perhaps some so rare that we have been hunting for

THE PROCUMBENT SIBBALDIA (*Sibbaldia procumbens*).

such a locality as this in the hope that we should
find them. This is the spot where the Scottish
Asphodel (*Tofieldia palustris*) is likely to be found.
In England it has been met with only in Teesdale,
but, as its name imports, it is not uncommon in
Scotland, as well as on the Irish mountains. It grows

to the height of six or eight inches, and bears a scape of yellow-tinted white flowers. It is a near relative of the Lancashire Asphodel (*Narthecium ossifragum*), which abounds in all the boggy places of the hill-sides in Wales and Northern England at much lower altitudes. A careful examination (and a *very* careful one is needed) by the sides of the miniature rills which trickle through this bit of bog may reward us by the discovery of the diminutive Bog Orchis (*Malaxis paludosa*), seldom more than a couple of inches in height. Note how its few soft leaves are fringed with little cellular bulbs, each of which is capable of developing a new plant. The yellowish-green flowers are only about one-sixth of an inch in length, but their structure, like that of orchids generally, is very ingenious.

In such places as these, especially on our northern mountains, we may be pretty sure of finding the Bog Whortleberry, or Great Bilberry, as it is also called (*Vaccinium uliginosum*). It ascends to between 3,000 and 4,000 feet, even in the Highlands. The drooping flowers are very small, and of a palish pink colour. The berries are smaller than those of the common Bilberry, in spite of the cognomen of "great." Perhaps we may come across the Cowberry (*Vaccinium Vitis Idæa*) hereabouts as well. It has a larger range in Britain than the Bog Whortleberry, although it is peculiar to mountain districts. The four-lobed flowers are of a pink colour, and are massed in terminal clusters. The fruit is red, and the leaves not unlike those of the Box. The Cranberry (*Oxycoccus palustris*) is another mountain marsh or

bog-loving creeping plant, common in the Highlands,
but abundant on the Irish mountains wherever the
Sphagnum moss extends. The flowers are very small
and of a reddish colour; the fruit is too well known,
both by sight and taste to need description. The
Bearberry (*Arctostaphylos alpina*) is a true Alpine
species, but **very rare,** being found only on a few of
the more barren Scottish mountains, from Perth and
Forfar northwards: its flowers are white. The
common Bearberry (*Arctostaphylos uva-ursi*) is much
more abundant, as it grows on most of our high
rocky places from the West Riding and Cumberland
mountains northwards, and it is also common on the
mountains of the north-west of Ireland. Its pink
vase-shaped flowers are $\frac{2}{3}$ inch in length, arranged
in crowded racemes on the depressed, trailing, but
stout woody branches. Let the student observe the
hairs which abound within the flowers of most of
these plants. They are for the purpose of keeping out
the small "unbidden guests," the feeble insects which
cannot cross the flowers. In Perthshire we should
find the lovely *Andromeda polifolia* in boggy places:
although it has a much more southerly range, it seems
to love the "bleak north" most. On the mountains
of north-western Ireland we should certainly meet
with an abundance of that most lovely of heaths,
Dabeocia polifolia, whose large rose-coloured bells
produce bright patches of colour on the lovely hill-
sides where it is massed. The first sight of the Irish
Dabeocia ("St. Dabeoc's Heath" it is also called,
after a notable Irish saint) produces in the mind of
the botanist a thrill of innocent delight. On the

bleak parts of the Sow of Athol, in Perthshire, we have the blue Dabeocia (*Dabeocia cærulea*). It is extremely rare,—indeed, perhaps one of the rarest members of our British flora. The moorlands of the lofty Scottish Alps, from Ben Lomond northwards, produce the Azalea (*Loiseleuria procumbens*). It grows in flat patches with interlaced woody branches, a habit it has doubtless acquired to enable it to be soon covered in the snow, and so protected from extreme cold. Many arctic woody plants possess this trailing character, and perhaps for a similar reason. The flowers of the *Loiseleuria* are not more than $\frac{1}{6}$ inch long, pink, and supported by a red calyx. The Winter-greens (*Pyrola*) are for the most part mountain plants, although some of them range as far as the south of England, and grow on heathy places at comparatively slight elevations above the sea-level. Such are *Pyrola minor* and *P. media;* yet we find both these species ascending to nearly 2,000 feet in the Highlands, so that we may regard them as sub-alpine species which have adapted themselves in some degree to changed conditions of climate. All the Pyrolas, however, have a very extensive geographical distribution. *Pyrola secunda* is found only in rocky mountain woods of the North, nowhere nearer south than the West Riding : it is a rare species. *Pyrola rotundifolia* is also a lover of moist woods and copses, and is of still rarer occurrence than the latter. *Pyrola uniflora* is almost entirely confined to the pine-woods of the northern Highlands, and is distinguished for the shorter size of the scape, and the solitary but larger flower. Generally speaking the flowers of all the Winter-greens are whitish, tinged sometimes with pink, and sometimes with green.

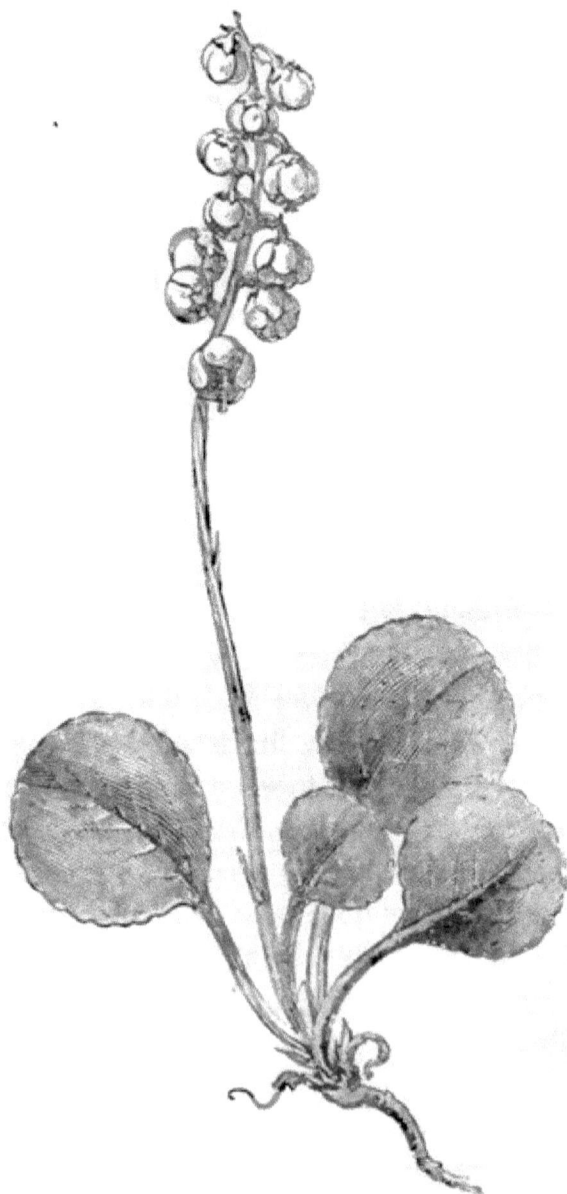

WINTER-GREEN (*Pyrola media*).

A very rare plant, allied to the Bearberry—
Epimedium alpinum, is
said to be native to
our mountains, although
Bentham thinks it is not.
The Epimedium is one
of those plants possess-
ing a very interesting
contrivance for prevent-
ing such small wingless
insects as ants from
climbing the stem and
flower-stalks, and thus
getting at the honey
within the flower. The
open structure of the
latter renders it impera-
tive that all insects shall

THE BARREN-WORT
(*Epimedium alpinum*).

approach it flying, for in this way they can benefi-
cially fertilize it. And so we find that, although the
lower parts of the stem and leaves of the Epimedium
are free from hairs, the flower-stalks are furnished
with them, of a kind which exude a gummy secretion.
These glandular hairs stand out horizontally, and
effectually prevent small insects from creeping up to
the flowers! It is in the elevated fir forests of midland and western Scotland that we meet with the
pretty little *Linnæa borealis* in an unquestionable
indigenous state. Every one is aware that the Father
of Botany adopted this flower (which was generically
named after him) as his favourite, and most of his
portraits show him with the *Linnæa* in his buttonhole.

The sweet-scented pink corollas are exquisite objects.
In Lapland it is one of the commonest objects growing,
the ground sometimes being literally crowded with it.
Trientalis Europæa is a pretty white or pale pink-
flowered plant, met with in mountain woods in the

Linnæa borealis.

North of England and Scotland : it is tolerably
common in the latter country.

The Cloudberry (*Rubus Chamæmorus*), although one
of the Brambles, is perhaps as characteristic a British
mountain plant as one could select. It is not un-

common on all our highest moorlands, ranging from
Derby and North Wales northerly into the Highlands;
although it is rare in Ireland. Its rootstock is creeping
and branched, and the short stem bears broad leaves
1 to 3 inches in diameter. The flowers are an inch
across, and of a lovely white colour, in appearance
like those of the common Bramble (than which no
flowers are more exquisite). The fruit is about half an
inch in diameter, and of an orange-yellow colour,
built up of a cluster of a few large " drupels," and
resembling the Dewberry in shape and size, although
not in colour. *Rubus saxatilis* is another well-known
creeping mountain bramble, longer than the Cloudberry, and bearing scarlet fruit : it is common in the
Highlands. *Rubus arcticus* is a beautiful little bramble,
which, however, has only been found in the Isle of
Mull.

The Crowberry (*Empetrum nigrum*) grows on the
heaths and rocks of the mountainous parts of England,
Ireland, and Scotland. In the latter country we find
it as high as 4,000 feet above the sea. Its slender,
wiry branches trail on the ground ; the leaves turn
red as they get old. The flowers are small and tend
to be " everlasting," owing to the scarious character
of the reflexed pink petals. This plant bears a red
eatable berry. The structure of its leaves is very
curious ; they form peculiar hollow cylinders, by
twisting over.

A pretty and much sought-after Alpine plant is the
Rose-root (*Sedum rhodiola*), one of the botanical rarities of Snowdon and other mountains. The marshy
localities of the hilly districts from York and West-

moreland northwards frequently yield a sub-alpine species of stonecrop (*Sedum villosum*), whose whitish purple flowers, although small, are conspicuous enough for the eye of the observer readily to detect it. The alpine Butterwort (*Pinguicula alpina*) grows also in moist places on the hills in the extreme north of Scotland. The alpine Bistort (*Polygonum viviparum*) would be less likely to be identified unless it is in flower, on account of its green grass-like leaves. It is partial to the moister mountain-pastures, although it may also be found growing on wet rocks at as great altitudes in Scotland as 4,000 feet. It is found on the Welsh mountains, particularly those above Carnarvon ; and on the higher hills of the West Riding. In Ireland, we believe it has only been found on the hills of the North-west. Its name of *viviparum* is given to this species on account of the little purple bulbs which are developed on the lower part of the flower-spike. Each of these takes root and becomes a perfect plant. Hooker tells us that the seeds or fruit of this species very rarely ripen, so that we have a compensation in the production of the bulbils, which prevent the plant becoming extinct. Damp sub-alpine meadows and banks are also the habitats of the alpine Scullcap

THE ROSE-ROOT
(*Sedum rhodiola*).

(*Bartsia alpina*), a species not uncommon on the hills and mountains of northern England, as well as on some of the Scottish mountains, notably those of Breadalbane, Inverness, and Ross; in places it may be found growing as high up as 3,000 feet. The plant is small, never more than 8 inches in height, and bears a spike on which are a few bluish-purple flowers, surrounded by purplish bracts. On the alpine moors of Yorkshire, the Cheviot hills, and in Scotland generally, we may look for the rare Swiss Dogwood (*Cornus Suecica*), having a low creeping rootstock, although the upright stem is not more than 8 inches in height. It bears an umbel of minute purplish flowers, which, if successful, ultimately yield small red berries.

The alpine Fleabane (*Erigeron alpinus*) is another and even a rarer upland flower, found on the Breadalbane and Clova mountains, at a height of 3,000 feet. It is a composite flower, exceedingly pretty, for the numerous ray-florets are purple, whilst the disk-florets are yellow. Each plant usually bears two or three flower-heads, about $\frac{3}{4}$ inch across.

In mountain woods we should look for a sub-alpine species of Cow-wheat (*Melampyrum sylvaticum*). Although rare, it is found in the woody copses which creep up the hill-sides from Durham to Westmoreland into Scotland. The small flowers are of a deep yellow colour. The alpine Enchanter's Nightshade (*Circæa alpina*) has a wide British range, for we find it (generally in the woods and copses of the hills and mountains) distributed over all our sub-alpine districts from Gloucester northwards. It is not un-

common in the mountain woods of the Lake district ;
and it is tolerably abundant in similar places in the
west and north-west of Ireland. It may be easily
recognised from the common Enchanter's Nightshade
by its leaves, which are much more deeply-toothed.
Kerner, in dwelling on the structures of plants for
the purpose of protecting them from what he aptly
calls "forbidden guests," says: "Among the numerous
small wingless creatures that frequent the place where
Circæa alpina grows, there are many of very small
size. Though these in visiting the flowers would
brush off the pollen, yet they are kept away by the
glandular hairs that stick out horizontally from the
lower and tubular part of the superior calyx, because
the pollen thus brushed off would be completely
wasted ; whereas small, flying dipterous insects, inasmuch as they flit from flower to flower in rapid suc-
cession, will carry the pollen with them, and transfer
it to the stigmas of other flowers."

In addition to the above-mentioned plants, we have
alpine representatives of genera which, in the
warmer climature and richer soils of the Lowlands,
attain the magnitude and dignity of forest trees.
Such are the Willows, Birch, &c., found on our mountain sides and summits. Nature has not been so
kindly generous to these alpine species as to their
temperate brethren below ; but it is astonishing how
they have adapted themselves to circumstances, nevertheless. The Willows, in particular, have several species
well fitted to live, in spite of the bleak conditions
of weather and barren soil frequent among the mountains. Of these perhaps the boldest and hardiest is

the Herbaceous Willow (*Salix herbacea*), which grows
on the summits of the loftiest mountains in North
Wales and the Lake district. It is still more common on all the high mountains of Scotland, where it
is met with at the great height of 4,300 feet. There
is a straggling look about it, however, and the stem
and branches spread horizontally under the turf, as if
hiding from the cold as much as possible. Here and
there they send up a few short, few-leaved flowering
twigs, which bear catkins not more than $\frac{1}{4}$ in. in
length. Unlike the habit of willows generally, this
species flowers *after* it has produced leaves. *Salix
reticulata* is another dwarfed and depressed species of
willow found only on the lofty mountains of Breadalbane, Clova, Braemar, and Sutherland ; but not at so
great a height as the preceding species. It may
easily be recognised from the latter by its much
larger catkins. The Lapland Willow (*Salix Lapponum*)
is not quite so dwarfed, perhaps because it never
ascends to such great altitudes. It is common among
the wet alpine rocks of Scotland. Sadler's Willow
(*Salix Sadleri*) is, we believe, known only to grow
(and there very rarely) on the rocky ledges of Glen
Callater, at a height 'of about 2,000 feet above the
sea-level. It is said that only *two* specimens of this
plant have as yet been seen. *Salix arbusculus* is another mountain willow confined to the mountains of
Scotland. *Salix myrsinites*, although chiefly Scottish,
is also found growing on the top of Muckish Mountains, in county Donegal. The Woolly Willow (*Salix
lanata*) is a very rare and very remarkable alpine
species. The broad ovate leaves are quite woolly or

cottony beneath : hence the specific name. It is
limited in its British distribution to the high cliffs,
and beside rocky rills in the Clova mountains, Glen
Callater, and Maol Cuachlan, in Scotland. *Salix
nigricans* is not found further south than York and
Durham; but at some height among the Scottish
mountains we find it growing to a length of 10 feet.
(We cannot say the *height*, because of its procumbent
habit.) The Creeping Willow (*Salix repens*) is fond
of heaths, moors, and hill-sides ; but it ranges far to
the south. It may, however, be regarded as a sub-
alpine plant, for it is found growing in the Highlands
at an altitude of between 2,000 and 3,000 feet.
Several other of our British willows, which we
have not space to mention, appear to be fonder of
moist hilly localities than any other, for we find that,
whether they spread southerly or not, they are sure
to be found among the hills when the circumstances
are favourable.

The Dwarf Birch (*Betula nana*) is another Alpine
plant peculiar to Scotland. Even there, however, it
is not abundant. It never grows to a greater height
than 3 feet. The Juniper (*Juniperus communis*) is a
better and more characteristic hill-side and mountain
shrub, abundant in the Lake district, where the rocky
surfaces of some of the smaller hills are frequently
miniature forests of it. In the Highlands it loves to
grow on the moraine-heaps of the upland valleys, and
on the lower flanks of the mountains, where its
numerous berries form an important element in the
larder of the Blackcock. It will be a good practice
for the young botanist to dissect the berries, and

compare them with those of the Yew, and the less attractive cone-seeds of the Pine and Fir. A well-marked variety of the Juniper, called *nana*, grows in the Highlands, at the height of 2,700 feet—300 feet

BRANCH OF JUNIPER, SHOWING MODE OF ATTACHMENT
OF BERRIES.

higher than the ordinary Juniper bushes can grow; so that it appears to be a variety adapted to extreme alpine conditions.

On the boggy parts of our moorlands we may

H

expect to find the Sweet Gale, or Bog Myrtle (*Myrica gale*), even at heights in Scotland of 1,800 feet. It is a shrubby, almost bushy plant, whose catkins

TWIG OR BRANCH OF SWEET GALE, SHOWING THE
FLOWERS.

usually flower before the leaves appear. The latter are exceedingly fragrant, or rather aromatic, owing to an essential oil which is secreted in the cells of the epidermis of the leaves. When the wind blows from the direction of a boggy thicket of Sweet Gale, we can perceive the fragrant odour even at a great distance. Cottagers and shepherds have long used the leaves for the purpose of making a kind of tea. The wood exudes a fragrant resin.

But, of course, the tree which we naturally associate with the mountains is that wrongly named a *Fir*, for it is a true *Pine*—the Scotch Fir (*Pinus sylvestris*). Not even the witches' tree, "Rowan," or Mountain-ash (*Pyrus aucuparia*), is more characteristic of the hills. Indeed, the latter is usually found in the gorges and glens, where its snake-like roots appear to cling to the naked faces of the cliffs, and its bright red berries offer a table in the wilderness to the mountain birds. The Scotch Fir loves to group itself in the most picturesque situations, where the sunlight can strike its rugged but warm-coloured bark, and the changing glaucous tints of its bunches of needle-shaped leaves can best be seen. Although now growing abundantly in Great Britain generally, the Scotch Fir is, as its name implies, naturally confined to the "Land of brown heath and shaggy wood." It has been carried all over England within the historic period, although, singularly enough, we have geological evidence of its being naturalized in this country before the commencement of the Glacial period, for its cones and branches are tolerably abundant in the submerged pre-Glacial "forest bed" of the

Norfolk coast. In the Highlands we find clumps of this noble tree at the height of 2,200 feet, although these are usually planted, and the tree itself is rare in the truly native state. Notice the peculiar difference in the male and female parts of this tree. The former are

BRANCH AND CONE OF SCOTCH FIR (*Pinus sylvestris*).

short, spiked, yellow catkins; and the latter cones of about two inches in length. Scotch Firs have been known to attain a height of 100 feet, and a girth of 12 feet.

We have a few other British Alpine and sub-Alpine

plants, not yet mentioned.   Some of them are ferns
and grasses.   But it will already have been seen that
the geographical distribution of Alpine plants over
high mountain flanks and summits, and the identity
of such plants occurring in these bleak situations with
those growing near or within the Arctic Circle, are
no mere coincidences.   No known theory will account
for the general distribution of Alpine plants in the
northern hemisphere, whether those of the Old or
the New World, other than that which we briefly
explained in the introduction to the last chapter.

CHAPTER VI.

THE FLORA OF OUR HILLSIDES AND MOORLANDS.

Alpine plants in the Lowlands—Alpine Ladies' Mantle—Common Ladies' Mantle—The Sneese-wort and Yarrow—The Field Gentian and Bog-bean — The Yellow-wort — Why Flowers "go to sleep"—Burns' Mountain Daisy—The Harebell, Eyebright, Milk-wort, Bird's-foot Trefoil, Sheep's Bit, Cud-weed, &c.—Wild Thyme and the reason for its Perfume. —Stone-crops — Burnet Saxifrage — Bilberry, its Various Names—The Gorse and its Various Names—Common Broom —Dyer's Green Weed—The Heaths and the Ling—The Sun-dews—Carnivorous or "Insectivorous" plants—The Butterwort—Professor Kerner on the Butterwort—Grass of Parnassus—Lancashire Asphodel—The Foxglove—"Doctrine of Signatures"—Ivy-leaved Campanula—Golden and Opposite-leaved Saxifrages—Welsh Poppy—Globe-flower—Rock-roses —Rock Hutchinsia—Red Centaury—The Yellow Mountain Violet.

EVERY botanist is aware that in addition to the special Alpine plants which have travelled from their arctic homes to our bleak mountain-peaks, there are others which own a less frigid origin. Since the Arctic species of plants took up their abodes in Britain, a few of them have been modified to bear a greater degree of warmth, and these we find have descended the mountains almost to their bases, and have mixed up, so to speak, with the southern flora which has taken possession of the plains. Still, such mountain plants have never spread far from the highlands. We seldom or never meet with them extending into the midland and southern lowlands of England. Such species as the Alpine Ladies' Mantle (*Alchemilla alpina*) is found in abundance at the feet of the Scottish mountains, mingling with the more temperate species known as the Common Ladies' Mantle (*Alchemilla vulgaris*). The latter we find further to the south, growing abundantly in the meadows of Cheshire, Nottinghamshire, and Worcestershire; although it is still a characteristic plant in hillside pastures, where its prettily crenated leaves and elegant form have obtained for it its popular name. This plant is an illustration of the curious fact that where the flowers are so intermixed with the leaves that destruction of the one involves destruction of the other, the leaves are avoided by grazing animals. The Common Ladies' Mantle is never eaten by cattle. In the southern and eastern counties the Field Ladies' Mantle (*Alchemilla arvensis*), with much smaller leaves, and altogether a less elegant and attractive plant, has taken its place. In this manner we frequently find that degrees of latitude might be represented by different species of the same genus of plant, one kind being better adapted to cold or moisture than another, as in these three British species of Ladies' Mantle. Any one who has wandered over the crisp hilly pastures of Lancashire, North Yorkshire, Westmoreland, and Cumberland, will have been particularly struck by the abundance of the Common Ladies' Mantle.

When we have three or more species of plants thus allocated to different districts, we shall usually find that each has its floral companions, so to speak.

Thus, with one characteristic northern hill-side plant,
the Common Ladies' Mantle, we shall find the Sneese-
wort (*Achillæa ptarmica*)—a species belonging to the
same genus as the better known and widely distributed Yarrow.   Its flower-heads are larger, and

THE SNEESE-WORT (*Achillæa ptarmica*).

therefore fewer than those of the latter plant.   Like
it, every part of the Sneese-wort is aromatic, only
more strongly so.   We have a yellow-flowered variety,
which is exceedingly rare.   Perhaps in the same hilly
or moorland pastures as the two plants just mentioned we shall find an abundance of the Field

Gentian (*Gentiana campestris*)—the "Baldmoyne" of old Chaucer's time. It is a branched, erect plant, bearing clusters of four-cleft flowers of a bluish-purple colour. In the centre of each we may note the fringe of hairs, which protect the nectar within from the depredations of small insects, whose visits could not possibly benefit the flowers, for they are too small to assist in "crossing" them by carrying pollen about; whilst, if they robbed the flowers of their nectar or honey, there would be then no inducement for larger and stronger and more beneficial insects to visit these flowers, and consequently the all-necessary "crossing" would not be brought about. Kerner has recently shown that nearly all the arrangements of minute hairs within the

THE FIELD GENTIAN
(*Gentiana campestris*).

interior of flowers have for their chief end the prohibition of too small and florally useless insects.

In addition to the practical usefulness of these floral hairs, they are frequently ornate additions to the attractiveness of the flowers. Any one noting how the hairy, almost lace-like fringe within the Gentian relieves the dull purple tints of their exterior will confess that it is a gain. The petals of the Bog-bean (*Menyanthes trifoliata*), which will be surely found in most quiet upland tarns and ponds in the northern counties, are covered with pinkish-white hairs, which renders it utterly impossible for too small and useless insects to walk over. Every one must assent that the Bog-bean is considerably indebted to these hairs for its peculiar loveliness. The Bog-bean is a near relative of the Gentian, and both are remarkable for the intense bitterness of the whole plant, but more especially of the leaves. In former times, ere hops had been imported, this species of Gentian was used by our forefathers for making their ale bitter.

If these hilly pastures be underlaid by limestone, we may be almost certain to find the brilliant Yellow-wort (*Chlora perfoliata*) growing in abundance. Its pointed egg-shaped leaves are covered with a sea-green "bloom." The flowers will be found to open at different times of the day, as if they relieved each other in keeping guard. The centre flower opens early in the morning, and closes at noon; this contrivance being evidently to evade the ants, which can easily climb the smooth stem and leaves, and get to the honey. But ants never stir abroad whilst the dew is on the ground or on vegetation. They wait until the sun has evaporated it. Meantime such flowers as the Yellow-wort, Goat's Beard, &c., are taking advantage of this circumstance, and are opening their flowers widely. But as soon as the day has grown, and the ants are out and diligently making up for late rising, then such flowers "go to sleep," as we term their early closing. Like the Gentian and the Bog-bean, which belong to the same natural order, the Yellow-wort has bitter leaves, and was formerly valued among country herbalists for its tonic properties.

Just as a few of the sub-alpine flowering plants spread tentatively into the lowlands, so do we find a few species from the plains gradually ascending higher up the hill-sides and over the moorlands. Among such we may mention the common daisy, which is dwarfed to a miserable poverty-stricken appearance high up the mountain-side. Burns, in his poem "To a Mountain Daisy," has hit upon the correct reason for this attenuated condition of a flower which in the warmer and more fertile meadows beneath blossoms in such luxuriance :—

> " Cauld blew the bitter-biting north
> Upon thy early, humble birth ;
> Yet cheerfully thou glinted forth
> Amid the storm,
> Scarce rear'd above the parent earth
> Thy tender form."

The Sheep's Sorrel, the Eyebright, the Hare-bell, the Dandelion, and many others, are frequently the companions of the Daisy in its moorland invasions. Of all flowers perhaps there are none which seem so much at home on our breezy heaths and lonely moorlands, or adorning the heather-clad mountain-slopes, as the common Hare-bell (*Campanula rotundi-*

folia). Many young botanists puzzle over its Latin
specific name, for they cannot find any *round* leaves
upon the plant. What few there are are linear, or
grass-like. But the fact is that the radicle leaves—
those which first appear above the ground—are round,
but they soon wither and die away. How exquisitely
graceful these flowers appear when the breeze wafts
them to and fro on their delicate hair-like stalks ! The
drooping condition of the bells is just that which enables
the rain to patter on the outside, and allows the dews
to trickle off harmlessly ; whereas if they were upright
they would soon be filled with such moisture, and all
their nectar be dissolved away, and the pollen-grains
spoiled ! In the heart of the Highlander this flower
is a rival even of the heather ! It seems so affec-
tionately related to all its natural surroundings that
we are not surprised the Lancashire poet, Edwin
Waugh, should have written of it so tenderly :—

> " And when it died the south wind sighed,
>   The drooping fern looked dim ;
>   The old crag moaned, the lone ash groaned,
>   The wild heath sang a hymn ;
>   The leaves crept near, though fallen and sere,
>   Like old friends mustering round ;
>   And a dew-drop fell from the heather-bell
>   Upon its burial-ground.

> " For it had bloomed, content to bless
>   Each thing that round it grew ;
>   And on its native wilderness
>   Its store of sweetness strew :
>   Fair link in nature's chain of love,
>   To noisy fame unknown,
>   There is a register above,
>   E'en when a flower is gone ! "

The common Eye-bright (*Euphrasia officinalis*) is one of our moorland gems. Its small, unequal-shaped white corolla, streaked with what Sir John Lubbock calls "honey-guides" (by which is meant the veins, &c., of petals, whose purpose is to direct insects to where the honey is stored away), and orange-yellow throat, all combine to render it one of the most elegant and beautiful of all our smaller flowers.

THE EYE-BRIGHT (*Euphrasia officinalis*).

To our thinking the Eye-bright always looks prettier on the hill or moor-side. It has fewer glaring rivals there, and we are forced to observe it closer. True, it does not grow to the same almost shrubby rankness which it attains in the lowland meadows. On the contrary here it is diminutive, but overladen with blossoms. Nestling amid the taller grass which grows under the shelter of some heather or gorse shrub, we shall be almost sure to find the wiry-stemmed lovely little Milk-wort. (*Polygala vulgaris*), or " Rogation

THE MILK-WORT (*Polygala vulgaris*).

Flower" of our ancestors, because it was used in Procession Week. The flowers (tassel-shaped) are usually of a sky-blue colour; but we may find white, and even the rarer pink varieties. The structure of this flower is well worth attention, being intended to store up pollen against the visits of insects, so that the hairy bodies of the latter may be dusted with it, and in this manner carried about from flower to flower. Singularly enough, this plant has always been regarded as causing cows to produce a large quantity of milk if they feed in the pastures where it grows. Both its popular English and scientific botanical names have reference to this very ancient opinion. It is surely worth while observing whether it be true or not, for there is frequently a substratum of wisdom in this kind of "folk-lore."

Higher up the hill-sides, where there is least moisture, and upon those parts of the moorlands where the short grass indicates dryness, we shall find growing intertwined among the herbage, the smallest and prettiest of our British leguminous plants, the Bird's-foot Trefoil (*Ornithopus perpusillus*), so called from the remarkable appearance which its three ripened seed-pods bear to the claws of a small bird's foot. The tiny blossoms are delicately pink and white, or cream-coloured, and the whole plant is recumbent, and the small pinnated leaves show it is allied to the Vetch family.

A reverent wanderer among these hills, looking out for these dainty little habitants of the springy turf, feels that all these floral subjects are—

> "Living preachers;
> Each cup a pulpit, every leaf a book;
> Supplying to his fancy numerous teachers
> From loneliest nook."

The heathy places are strewn everywhere with the
Sheep's Scabious (*Jasione montana*), or "Sheep's Bit,"
as it is also called, said to be a favourite morsel with
the woolly tribes whose name it bears. It is of a

THE BIRD'S-FOOT TREFOIL (*Ornithopus perpusillus*).

pretty light-blue colour, and its abundant flowers frequently make the dry heather glad by their presence.
Not far from its neighbourhood (for both love the
dry heath soils of our moorlands) we shall come
across the cottony Cud-weed (*Filago Germanica*), a

composite plant allied to the "Everlasting flowers" of our gardens. The quaint old botanists gave the name of "*Herba impia*" to this plant, on account of the branching flower-stalks, or young shoots, being higher than the parent flowers, and therefore treating the latter with disrespect by over-topping them !

These dry, short, heathy pastures are in the summer-time often perfect sheets of colour. The Wild Thyme, and Yellow and White Stone-crops completely mat the ground in variegated patches; whilst the trailing stems and little white flowers of the Stone Goosegrass (*Galium saxatile*) and the trailing St. John's Wort (*Hypericum humifusum*) are interweaved among the contrasted pat-terns. The Wild Thyme (*Thymus serpyllum*) is too well known to need descrip-tion. Its fragrance is always most powerfully emitted when the summer sun shines

SHEEP'S SCABIOUS
(*Jasione montana*).

fiercest, and as perfumes have the power of barring out heat-rays, there can be no question that this quality prevents the thyme from being completely scorched up on the dry soils where it most loves to

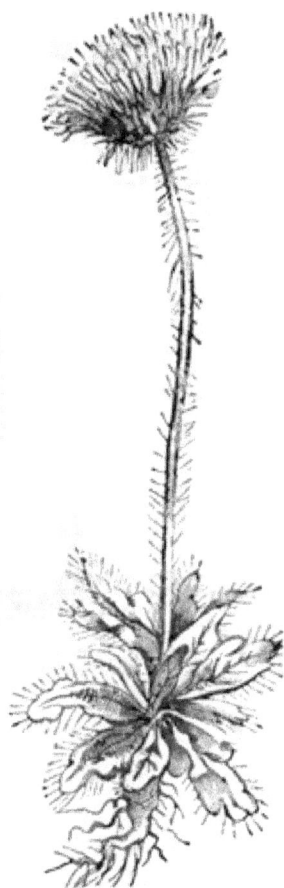

grow. Moreover, its presence may be protective to other plants, which reap the benefit, during hot weather, of its thermally opaque perfume. Again, we know that this aromatic perfume attracts its flower-frequenting insects, and the well-known "Honey of Hymettus" is said to have owed its peculiar fragrance to the wild thyme flowers whence the bees had gathered it. Anyhow, these tracts of heath or moorland dry pastures and banks—"whereon wild thyme grows"—are perfect metropolises of insect life. The air is full of their busy and varied hum. Shelley speaks of "the bees on the bells of thyme," although in reality the flowers are not bell-shaped, but lipped (*labiatæ*). Among our English poets the fragrance of this plant has long been proverbial—

> " Airy downs and gentle hills,
>   O'er grass with thyme bespread."

Armstrong speaks of these healthy spots where

> " Thyme, the loved of bees, perfumes the air."

As you walk over the hilly greensward the foot crushes the diminutive but wiry little plant, and causes it to exhale its aromas all the more powerfully. The Yellow Stone-crop (*Sedum acre*), which abounds in such dense golden masses in June and July, can do with little moisture or soil, as indeed can all the family to which it belongs. Notice how the House-Leek thrives from year to year on the bare slates or tiles of the cottagers' roofs. The thick, fleshy leaves of the Biting Stonecrop are very acrid to the taste, and as it grows over waste stony places where few other plants can, it has earned the name of "Wall Pepper," from its pungency. It is a favourite plant to grow over garden rock-work, for when it has not burst out into a mass of yellow blossoms, the greenness of its creeping stems (which assume a new freshness after every rainfall) upholster the rugged *omnium gatherum*

THE BITING STONE-CROP (*Sedum acre*).

which usually compose such structures. The White English Stone-crop (*Sedum Anglicum*) is not un-frequent on moorlands and heaths adjacent to the sea; although the White Stone-crop (*Sedum album*) is the species which usually takes duty with the yellow. Wallace, than whom we have no more en-thusiastic naturalist living, after years' experience in tropical countries, amid their flowers and insects, tells us that in all his travels he has seen no sight which for floral colour could equal an English meadow in May! We might almost say the same of these heath-clad slopes.

It is on these dry pastures only that we shall find a characteristic umbelliferous plant, the common Burnet Saxifrage (*Pimpinella saxifraga*).   It grows to the height of one or two feet, its lower, sharply cut pinnated leaves being grouped on long stalks.   The upper leaves are bi-pinnate, and deeply cut into sharp, narrow segments.

But it is unquestionably the dense growth of such shrubby plants as the Bilberry, Gorse, Broom, and different species of heath, which gives to our uplands their most charming floral characters.   In the summer time they light up the hillsides and moorlands with their bright colours, whilst in the winter their dark masses throw a wild gloom over the landscape where they grow.   The Bilberry (*Vaccinium myrtillus*) is also called Whortleberry in the West of England; and in Lancashire (where its sloe-like fruit is in great request among the poor people) it is known as the "Whinberry," from its growing where the whin or gorse bushes are abundant.   Its nearly globular, waxen flowers are green, just tinged with purple, and these give rise to delicious black berries, whose juice stains the mouth when eaten.

Rugged, spiky, tortuous though it be, our English Gorse (*Ulex Europæus*) is without doubt the characteristic habitant of our heaths and moorlands.   In different books we have the story told, sometimes of

Linnæus, sometimes of Dillenius, that when he first
visited England and saw the commons sheeted over

THE WHINBERRY OR WHORTLEBERRY (*Vaccinium myrtillus*).

with the yellow blossoms of this plant, he fell on his
knees in grateful delight! It does not much matter
which of these two celebrated botanists thus gave

way to his feelings—the story has a purpose, if only to show that the gorse has a bold beauty of its own.

In the South of England this shrub is usually known as "furze"; in the eastern and north-eastern counties, as well as in parts of Lancashire, as the "whin," and in the north as the "gorse." Perhaps the latter English name is that by which it is most widely known. The young and tender shoots, ere they stiffen and harden into the well-known spines, are excellent "feed" for sheep, and a gorse-common is therefore a good feeding-ground for them. Nearly the whole year round this bush shows some of its yellow blossoms; hence the old proverb that "when the gorse is out of bloom, kisses are out of season." It is the story of the great Swedish botanist, expressing his sense of the glorious beauty of this plant, which Mrs. Elizabeth Barrett Browning has enshrined :—

> " Mountain Gorses, since Linnæus
>   Knelt beside you on the sod,
>   For your beauty thanking God,—
> For your teaching ye should see us
>   Bowing in prostration new."

And the same gifted lady alludes to the nearly perpetual flowering of the Gorse (although May is the month when it bursts forth in all its floral strength), in the same poem :—

> " Mountain blossoms, shining blossoms,
>   Do ye teach us to be glad
>   When no summer can be had,
> Blooming in our inward bosoms?
>   Ye whom God preserveth still,
>   Set as lights upon a hill,
> Tokens to the wintry earth that Beauty liveth still.

"Mountain Gorses, do ye teach us,
    From that academic chair
    Canopied with azure air,
  That the wisest word man reaches
  Is the humblest he can speak?
  Ye who live on mountain peak,
Yet live low on the ground, besides the grasses meek."

All the livelong summer day—and the hotter it is
the more one will hear it—the dry seed-pods of the
Gorse are continually exploding, now in volleys, now
in sharp fusillades or peppering shots. A small and
much rarer species of Gorse (*Ulex nana*) is sought
after by botanists, and esteemed a "good find."

Belonging to the same natural order is an allied
shrub, not uncommon on many hillsides, where its
lighter yellow blossoms mass together almost as
vividly as those of the Gorse—the common Broom
(*Cytisus scoparius*). It ranges from Caithness southwards, and is found in the Highlands at as great a
height as 2,000 feet. Everybody is acquainted with
the appearance of this beautiful shrub. In former
times its young shoots were much sought after for
diseases of the bladder and dropsy; and in the
northern counties and Scotland its medicinal credit is
still great among the common people. One of the
old Scottish ballads thus speaks of this familiar
plant :—

"O, the Broom, the bonny, bonny Broom,
    The Broom of the Cowden knowes ;
  For sure so soft, so sweet a bloom
    Elsewhere there never grows."

And Burns mentions it in one of those delicious

poems which so graphically bring before us the principal flowers of the heaths and moorlands :—

> " Their groves of sweet myrtle let foreign lands reckon,
>   Whose bright beaming summers exalt the perfume ;
> Far dearer to me, yon lone glen of green breckon,
>   Wi' the burn stealing under the long yellow Broom.

> " Far dearer to me are yon humble Broom bowers,
>   Where the bluebell and gowan lurk lowly, unseen ;
> And where, lightly tripping amang the sweet flowers,
>   A-listening the linnet, oft wanders my Jean !"

The Gorse as well as the Broom gives out a pleasant
perfume, although some people do not enjoy it. One
of our poets speaks of

> " The golden Broom
> Which scents the passing gale."

Nearly allied to the Broom is another plant which
we have frequently gathered on the hillsides of
Cumberland and Westmoreland—the Dyer's Green-
weed (*Genista tinctoria*). Its popular name originates
from its habit of turning a bright green colour in
drying. For generations this plant has been used for
a yellow dye it yields, and in connection with the
plant anciently known as "woad" for dyeing wool a
green colour. It was a species of *Genista*, and not
the common Broom (*Planta genista*) which the Plan-
tagenets adopted for their knightly badge, and which
afterwards gave them their family name.

But we have hundreds of square miles of hillside
heath and moorland which know little or nothing of
the shrubs just mentioned, and where the different
kinds of heather have the monopoly of the ground.

The two species which more or less abound on every moor are, first, the common or Fine-leaved Heather (*Erica cinerea*), and next in degree the much prettier Cross-leaved Heath (*Erica tetralix*). Neither is so

FINE-LEAVED HEATH (*Erica cinerea*).

abundant, however, as the Ling (*Calluna vulgaris*). The first-mentioned species has its leaves in *threes*, and the vase-shaped flowers are in irregular-whorled clusters. The Cross-leaved Heath has the leaves in *fours*, whilst the larger blush-tinted flowers cluster

COMMON LING (*Calluna vulgaris*).

more on the summits of the young shoots of the wiry stalks. The Fine-leaved Heath (*E. cinerea*) was formerly used in brewing beer. The Highlanders have long employed it for dyeing their tartans and making their beds. It is the Ling, however, to which the name of *Heather* is distinctively applied. This plant grows to a greater size than the afore-mentioned heaths, not unfrequently growing to the height of two and three feet on the Highland mountains. This it is which gives the moorland landscape its chief botanical feature, and whose purple flowers light up the otherwise dreary hillsides with a glow which in a picture seems too unreal. The Ling it is on whose seeds the grouse and blackcock feed, and which affords these moorland birds a winter shelter. From its tiny blossoms the bees for miles around derive

THE CROSS-LEAVED HEATH
(*Erica tetralix*).

their honey, and to the naturalist these waste places are full of busy industry and animal life. The botanist is aware that the mouths of these heather flowers are so constructed, partly by the

closing-in of the bell-shaped corolla and partly by the
clustering arrangement of the eight stamens just
within, that only certain kinds of insects, such as
bees, can obtain the honey inside, smaller and useless
insects being in this manner kept out. Not unfrequently you may see scores of heather bushes interlaced by the common Dodder.

We have no other combination of British flowers
which gives one such a sense of freedom as the wild
fragrance of a heath or moor in full heather bloom !
Eliza Cook's lines express the feelings of many who
love the breezy hills :—

" Wild blossoms of the moorland, ye are very dear to me ;
  Ye lure my dreaming memory as clover does the bee ;
  Ye bring back all my childhood loved, when freedom, joy
     and health
  Had never thought of wearing chains to fetter fame and
     wealth.
  Wild blossoms of the common land, brave tenants of the earth,
  Your breathings were among the first that helped my spirit's
     birth ;
  For how my busy brain would dream, and how my heart
     would burn,
  Where Gorse and Heather flung their arms above the forest
     fern."

We get rarer species of Heath (*Erica Mackaii, E.
Mediterranea, & E. ciliaris*) in the West of Ireland,
and right worthy are these lovely plants of the botanist's wanderings in search of them. Cornwall also
yields us a special form from its extensive heaths, the
Erica vagans, with rose-coloured flowers. The chief
locality for the latter plant is the Goonhilley Downs,
where there are whole acres of it, almost to the
exclusion of other plants.

It is on the margins of dry heathery banks, and in the vacant places between gorse-bush or ling, that the Ladies' Yellow Bedstraw (*Galium verum*) loves to grow. Its upright spikes of canary-coloured flowers are well-known objects ; as is also the perfume, not unlike that of hay, which the flowers plentifully emit. The Bedstraws were originally written " Bead-" straws, and the adjectival name of Lady was one of the many ancient references to the Virgin Mary, given in pre-Reformation times. " Ladies' Bedstraw" therefore really means "Our Lady's Bead-straw," the "beads" being the clustering, round, bead-like seed-vessels borne by these plants.

These dry heathery soils produce the yellow flowers of the common Tormentil (*Potentilla tormentilla*) and the creeping Cinque-foil (*Potentilla reptans*) in greater abun-

THE CORNISH HEATH
(*Erica vagans*).

dance than we shall find them growing elsewhere. The former may readily be recognised by its *three*-fingered leaves, and the latter by its having *five*. Both plants are sought after by the moorland herbalist on account of their supposed febrifugal and other virtues.

Let us now make our way to the bit of bog, whose patch of different colour and freedom from bushes, has made it conspicuous.   The grass grows ranker and

THE ROUND-LEAVED SUN-DEW (*Drosera rotundifolia*).

longer around it, and there are square yards of surface looking quite velvety with the yellowish-green Bog-moss or *Sphagnum*.   This is just the place to look for that now most interesting plant,   the Sun-dew

(*Drosera rotundifolia*), and many other moisture-loving "bog" plants, such as the Marsh St. John's Wort (*Hypericum elodes*), with its yellow flowers and hairy stems and leaves; the Bog-Pimpernel (*Anagalla tenella*), with its small, wiry, pinnated leaves and lilac-tinted corolla; the Lancashire Asphodel, Butterwort, and lovely Grass of Parnassus. We may perhaps find the long-leaved, as well as the round-leaved species of Sun-dew. Both these plants derive their popular names from the tiny dew-like drops secreted by the myriads of red, glandular hairs with which the upper surfaces of the leaves are crowded. We know (from the researches of Darwin, Hooker, and others) that these Sun-dews occupy a prominent place among that remarkable group of plants to which the name of "Carnivorous," or "Insectivorous," has just been given, from their habit of feeding on and consuming dead insects. The "dew," secreted by these red hairs, is very sticky, and small insects alighting on the leaves are caught, just as birds are by traps set with birdlime. Moreover, the red hairs have the power of then individually bending over the wretched insects until they are strangled as well as limed, even the edges of the leaf curling over and assisting in the strangulation. As soon as the insect is dead, and as fast as it decomposes, its tissues are absorbed by the leaves, which secrete a juice capable of partially digesting them. When an insect is thus digested, the leaves uncurl, the glandular hairs reassume their erect position, and the "dew" exudes from their tips—in short, the trap is set for another victim! You cannot examine a single plant of the Sun-dew without finding various insects entangled in the leaves. Note how the entire plant employs its thread-like roots only for the purpose of *anchoring* itself to the upper surface of the layer of *Sphagnum* moss. The roots have thrown all the duty of obtaining nitrogenous substances on the leaves, and in turn now assume that of anchoring the plant whilst it fulfils its strange and anomalous functions. The Sun-dews bear pretty but inconspicuous spikes of white flowers.

The Butterwort (*Pinguicula vulgaris*) is another bog-plant, or one only to be met with in perpetually damp situations. Its common name expresses the characteristic *greasiness* of its pale green leaves, whose upturned edges enable them the better to carry out their functions of "fly-catching." For the Butterwort is also "insectivorous," marsh insects being retained by the greasy exudation, which is at once their trap and the liquid for digesting them.

Kerner has recently shown that the greasiness of the leaves of the Butterwort performs another, and possibly even a more important function than that of occasionally entangling flies. There can be little question that the leaves in reality do double duty. We will listen to what this original botanist says himself :—" The structure of these flowers (Butterworts) reminds one very much of the Bromeliaceæ, where the rosette of leaves forms a basin, out of the middle of which rises a slender flower-stem." Prof. Kerner had previously shown that this method of isolating a flower-stem prevented creeping insects from climbing and getting at the honey, exactly on the principle adopted in some countries of protecting articles by placing them on tables whose legs are made to stand in water. The writer goes on to say, "but whereas in the Bromeliaceæ the basin is filled with rain and dew, in several species of *Pinguiculæ* the upper surface of the leaves which form the rosette is covered over with a tenacious, viscid slime. This sticky matter is secreted by small glandular trichomes, that are so thickly crowded together on the upper surface of the leaves that in *Pinguicula alpina* I could count nearly one hundred of them on a square millimètre. These glandular trichomes are of two kinds. The secretion which is discharged by these trichomes is colourless, slimy, and very tenacious. No small animal that comes in contact with it, and once adheres, can ever get free. The largest insect which I found sticking to this viscid layer was a dead ant (*Myrmica lævinodes*), an ant four millimètres in length. Should any larger and stronger insects get on to the leaf-rosette, they can manage to free themselves from the viscid substance. When they have accomplished this, they always try to reach the outer edge of the rosette, so as again to get firm ground under their feet, and they avoid climbing up the flower-stalk which rises nearly from the centre of the rosette. Darwin's statement that the glandular trichomes on the upper side of the leaves are stimulated to increased secretion by contact with these insects, firmly stuck, or I might almost say, imbedded in the slime, and that the insects themselves are actually digested, I can simply confirm by my own observations. Neither can there be a doubt that the dissolved nitrogenous constituents of these insects are absorbed by the

plants and utilized as nutriment; but no less certain
is it that four species of the *Pinguiculæ* which I have
investigated flourish perfectly well without animal
food, and therefore are not dependent on it. The
primary function of the glandular trichomes on the
leaves of *Pinguiculæ* and numerous other plants is
certainly, therefore, to keep off those creeping insects whose bodily dimensions are so small that
their visits would not bring about crossing; but
this, of course, does not exclude the possibility of
such insects as get caught and remain adherent being
digested, and serving as welcome, if not very luxu-
rious food."

The Butterworts bear each a blue violet-shaped
flower, on a single long stalk. The Grass of Parnassus
(*Parnassia palustris*) indicates its presence by its
lovely white flowers, each petal of which is streaked
with light green "honey-guides." At last Kerner has
shown the meaning of the five bunches of glands
which are an ornament, as well as use, to the interior
of the flower. They are for the purpose of protecting
the nectaries from the attacks or ravages of small and
(to the plant) useless insects. The elegantly-shaped
and bright green leaves of the Grass of Parnassus will
mark its habitat, even when the lovely white flowers
are not present.

It is in such boggy places as these, or on the lower
slopes of the hills and mountains where there is
always a good supply of moisture from superficial
drainage, that we shall find the Lancashire Asphodel
(*Narthecium ossifragum*) growing in abundance. It is
a lovely yellow plant, with vermilion-coloured stamens.

But pretty and attractive as it is, it is only one of the
Rushes which have been
favoured by fortune with
colour. Compare it with
the common Field Rush,
and you will see that their
flowers agree in every respect, except that the last-
mentioned is colourless.
For generations this plant
was much dreaded by moor-
land and hillside farmers,
who firmly believed that if
their sheep partook of it,
their bones would gradually
rot! The botanical specific
name of *ossifragum* still en-
shrines this curious piece
of superstition.

It may be that in our
moorland ramble we have
come to the natural out-
crop of the grit rock,
along whose face the damp
moisture is exuding, draining away beneath the copse
below in a series of tiny
rills. Examine the stones
hereabout, for sheltering
under their edges you may
find that diminutive but

THE LANCASHIRE ASPHODEL
(*Narthecium ossifragum*).

most elegant of our British plants, the Ivy-leaved

Campanula (*Campanula hederifolia*). In just such localities have we found it in North Wales, particularly near Capel Curig; and on Dartmoor, in Devonshire. Where the soil in the copse has been washed down

THE IVY-LEAVED CAMPANULA (*Campanula hederifolia*).

and is constantly moist, the ground may be covered with the delicate lovely Golden Saxifrage (*Chrysosplenium oppositifolium*). Its foliage is inexpressibly tender in its golden green tints, which are quite as attractive as the flat clusters of yellowish-green flowers which terminate them. You should visit the damp woods and rocks late in April to see the Golden Saxifrage in all its glory. Another species (*C. alterni-folium*) differs from the above commoner form by the arrangement of the leaves.

Here too (if on grit rocks) we may expect to find a grove of that most magnificent of all European plants, the Foxglove (*Digitalis purpurea*). It rises

THE GOLDEN SAXIFRAGE (*Chrysosplenium oppositifolium*).

majestically from its dense cushion of hairy leaves, and presents its flowers in such a way as to invite insect visitations. Note the peculiar arrangement of the stamens within the "gloves." Lubbock tells us that these flowers are "exclusively fertilized by humble-bees, which alone are large enough to fill the bell, and thus to deposit pollen on the stigma." But if the visits of bees are delayed or prevented, singularly enough there is an arrangement for rendering

the plant self-fertile ! Both in France and Germany
the popular name for this widely-diffused plant is
" Finger-flower " ; and, although objections have been
raised to our interpretation of its common British
name, we cannot but think that it originally meant ,
" Folk's glove "—that is, the glove of the fairies, or
" good folks." Note the German similarity, in its em-
ployment of the word " finger." An old poet, William
Browne, thus describes Pan as seeking gloves for his
mistress :—

> " To keep her slender fingers from the sunne,
> Pan through the pastures oftentimes hath runne
> To pluck the speckled Foxgloves from their stem,
> And on those fingers neatly placed them."

in which we get an indirect allusion to the origin
of the common name, as above interpreted.

The Foxglove is still largely used in medicine,
especially in the Pharmacopœia of the homœopathists.
Old country herbalists still believe it is a remedy for
ailments of the lungs, and this idea appears to have
descended from the time when the " doctrine of
signatures " was an article of undisturbed botanical
faith. Then it was imagined that every plant was
good for some human ailment or another, if we could
only find it out, and that many plants bore some out-
ward and visible sign, in colour or marking, of what
they were good for. Thus the spotted leaves of the
Lungwort and flowers of the Foxglove, it was thought,
showed they were curative of pulmonary complaints.
Nobody ever thought it was possible that flowers had
been created *without* reference to man ! If modern
science has done nothing more than destroy that egre-

gious human self-conceit which held that nothing had a right to live that was not of service to mankind—it has rendered good service. A false and spurious

WELSH POPPY (*Meconopsis Cambrica*).

religious sentiment ran through the old doctrines, and perhaps for a long time kept them from intellectual decay. Thus, regarding the medicinal virtue of the plant in question, a writer says :—

" The Foxglove leaves, with caution given,
      Another proof of favouring Heaven
      Will happily display :
  The rabid pulse it can abate,
  The hectic flush can moderate,
  And, blest by Him whose will is fate,
      May give a lengthened day."

On the damp slopes of the Welsh hills, here and there in sheltered places, but much more abundantly on those of Westmoreland and Cumberland, we may meet with the Welsh Poppy (*Meconopsis Cambrica*) and the noble Globe-flower (*Trollius Europœa*). Nowhere have we seen these remarkable plants growing in such profusion as on the Lancashire side of Lake Windermere. The latter is a well-known garden plant, where it exhibits the usual tendency to become double, through the stamens being converted into petals. It is in reality a sub-alpine flower, and in Scotland, where it is known as the " Lucky Gowan," grows as high up as 3,300 feet. Nobody will mistake the Welsh Poppy for the class to which it belongs, for in its large, flabby, yellow petals it strongly resembles the well-known yellow Horned Poppy of our sea-sides. In Scotland, however, this plant is said to be naturalized.

If the hill-sides be formed of limestone, we shall fail to find the Foxglove, for it has a singular dislike to calcareous soil. On the other hand, we shall come across plants we did not see elsewhere, such as the Yellow Corydalis (*Corydalis lutea*), whose pale-green leaves so much resemble in size and appearance the Maiden-hair Fern that the young botanist may imagine he has stumbled across a remarkable " find."

It loves the damp, stony places of the limestone hill-sides, and in the Peak of Derbyshire, as well as in North Lancashire and West Yorkshire, where the car-

THE GLOBE-FLOWER (*Trollius Europæa*).

boniferous limestone occurs, we have seen it covering the rock-surfaces for yards together. On the drier parts of these limestone rocks, also, we should look out for two or three species of the lovely Cistus, or Rock-rose (*Helianthemum*), and it will perhaps be found as high up as 2,000 feet above the level of the sea. *Helianthemum vulgare* and *H. canum* are most likely to be the species found. The flowers of the former are of a canary yellow, the petals lax, and the entire blossom about an inch in diameter. *H. canum* is rather rare in its occurrence; but even more than the former, it will be found at considerably high elevations. The flowers are much smaller than those of the common species. We have both species abundant on the limestone hills of Derbyshire; where also grows, at places in considerable abundance, a somewhat rare cruciferous plant called *Hutchinsia petræa*. It is also found under similar

HOARY ROCK-ROSE
(*Helianthemum canum*).

situations in North Wales and West Yorkshire, up to
the height of 1,500 feet. Its botanical name was

COMMON ROCK-ROSE (*Helianthemum vulgare*).

given to it in honour of Mr. Hutchins, a well known
Irish botanist.

It is on the dry hill-sides also that we shall find
the Red Centaury, or "Sanctuary" as the Lancashire
and Yorkshire herbalists term it. It is a great
favourite with the latter class for its many supposed
virtues, some of which are doubtless real, for its

THE ROCK HUTCHINSIA (*Hutchinsia petræa*).

bitterness makes it a good tonic. But its exceedingly
pretty appearance cannot fail to attract the attention
of those who are interested in the wild flowers of our
hills and moorlands. Its flowers are of an attractive
rose-colour, with yellow throats. The flowers always
close in damp weather, when the plant might be

passed over unheeded. The Yellow Mountain Violet (*Viola lutea*), perhaps the loveliest of our wild species, is abundantly distributed along the flanks of the Derbyshire hills.

A moorland ramble after these various wild flowers will give the pedestrian the opportunity of noting the habits of many birds and insects which resort to the quietness of these upland solitudes. Such rambles linger long in the memory, and are cherished as among the most elevating of our summer recreations. We are brought face to face with Nature and Nature's God! Our hearts instinctively go forth with new reverence for His creatures; and we feel with a gladness and faith which words fail to express, that " His loving-kindness is over all His works!"

CHAPTER VII.

THE RUSHES, SEDGES, AND GRASSES OF OUR MOUNTAINS AND MOORS.

Plants adapted to variety of Physical Conditions—Anemophilous
or Wind-fertilized Plants—The Cotton Grasses—Attempt to
Utilize them—Sedges in Moorland Bogs—Flea Sedge, Mud
Sedge, Creeping Sedge, Alpine Sedge, Black Sedge, Rock
Sedge, Mountain Sedge, &c. — Mountain Rushes — Alpine
Curved Rush, Spiked Wood Rush, Clustered Rush, Three-
leaved Rush, &c.—Mountain Grasses—Mat-grass, Wavy
Mountain Hair-grass, Alpine Hair-grass, Sheep's Fescue-
grass, &c. — *Poa alpina* — Alpine Foxtail-grass — Alpine
Timothy-grass, Oat-grass, &c.—Very rare British Mountain
Grasses — Blue Moor-grass, Heath-grass, &c. — Grasses
adapted to Elevated Conditions.

FROM the broad and easily-comprehended generaliza-
tion which philosophical botanists have recently made,
to the effect that all attractive, coloured, or perfumed
flowers are crossed by insects, whilst all the inconspicuous, uncoloured, and unperfumed flowers are
crossed by the agency of the wind, it would be expected beforehand that the breezy moorland slopes
and mountain-summits would be most favourable to
the growth of wind-fertilized plants. And such we
find is actually the case. Every physical condition
of these uplands—boggy, wet, dry, and rocky—has its
species of rush, sedge, or grass, which appears to
have selected the spot as the best and most favourable

to its development. In numerous instances we find plants belonging to those natural orders which cannot possibly live in the plains. They are even more Alpine and sub-Alpine than the greater part of the mountain flowers we have already drawn the attention of the reader to.

The flowering parts of grasses, rushes, and sedges are so inconspicuous—that is to say, they so rarely possess any of those attractive qualities with which the popular mind has learned to associate flowers— that many unscientific people are surprised to find them grouped among the true flowering plants. But flowers they are, nevertheless, with stamens and pistils, each re-acting on the other. Note the feathery and oftentimes branching character of the pistils of any of these plants. Could any floral mechanism be better contrived to catch the stray pollen-grains which the wind blows about ? And the anthers, or pollen-bags, observe how they are suspended on long filaments, whose sudden growth lifts them, so to speak, out of the chaffy glumes (the analogues of those very parts which in the Lilies are so gorgeously coloured) which have so well protected these tender reproductive parts whilst they were being developed ! And now that the stamens are thus enabled to dangle outside, is it not evidently an ingenious contrivance for enabling the wind to catch up and carry away the pollen-grains as fast as they are set free from the ripened *anthers* or pollen-bags?

We have already referred to the Tofieldia and Lancashire Asphodel (*Narthecium*) as being rushes, in spite of the beautiful appearance of the flowers of

the latter. Some botanists arrange them with the
Lilies on this account. The only charm which the
mountain grasses, &c., have for us is their elegant
and sometimes graceful shapes. The margins of the
mountain lakes and tarns, and the boggy marshes
formed where the surplus
water oozes to a lower
level, are the best places
to look for the sedges in
particular. Among the
most striking of these
moisture-loving species is
unquestionably the
Cotton-grass (*Eriopho-
rum*), whose white, silky,
cotton-like heads give the
hill-sides a peculiar ap-
pearance. Most people
who ascend mountains
and wander over moors
have noticed this plant,
and not a few practical
minds have asked whether
it could not be turned to
some useful account. If
patents for the utilization
of the common cotton-
grass have not been taken out, we know for certain
that intentions to do so have been entertained. But
hitherto no result has been attained, nor do we think
any practical *remunerative* results are likely to ensue
as long as the true Cotton-plant (*Gossypium*) can be

THE TASSEL COTTON-GRASS
(*Eriophorum gracile*).

so cheaply cultivated ; apart from the difficulty which the straight and stiff hairs of the Cotton-grass present.

It is when the Cotton-grass is in *fruit*, not when it is in *bloom*, that it assumes this attractive appearance. The rarest of the species is *E. alpinum*, which has been lost in one mountain bog (that of Restennet, Forfar), where it formerly abounded, through being drained. The "cotton" produced by the narrow-leaved species (*E. angustifolium*) appears most likely to be of use, and indifferent cloth has actually been woven from the thread spun from the hairs of this plant.

The botanist soon discovers that the mountain grasses and sedges form a very large proportion of our native Alpine plants. And perhaps no other group has been so largely adapted to climatal conditions. On the upland moors of the West Riding, Westmoreland, and many parts of Scotland, in the dampest places, we should look out for the tolerably rare sedge *Kobresia caricina*, a very rigid species, leafy only at the base. The bogs in mountainous districts are capital hunting-grounds · for species of sedges we shall probably not find elsewhere. Here, a careful search may present us with the graceful Few-flowered Sedge (*Carex pauciflora*). It is most abundant on the Scotch moors, where we find it at nearly 3,000 feet, and comparatively rare elsewhere, except perhaps in Yorkshire. The singular-looking Flea Sedge (*C. pulicaris*) is much more abundant in boggy situations ; as is also the Little Prickly Sedge (*C. stellulata*). We find the latter common on the boggy parts of the Highlands, at an altitude of 3,000 feet. Over our northern moors, and still more commonly distributed in the Highlands and the mountains of Sutherland, wherever there is marsh or bog, is the Creeping Sedge (*C. dioica*), remarkable for the male and female flower-spikes being separate ; that is to say, one bears stamens only, and the other pistils. A more local sedge, as regards its occurrence, is the Loose-flowered Mud Sedge (*C. rariflora*). It abounds on the elevated, boggy table, and between the counties of Aberdeen and Forfar, at heights of close on 3,000 feet. This is a peculiarly Arctic or Alpine species ; quite as much so as the Alpine Sedge (*C. alpina*), which latter, however, is a much rarer kind, and one that grows in quite a different situation. The Alpine Sedge loves to grow where we may sometimes find several other mountain species—on the damp ledges of rocks, at the bases, or midway up the crags. Hence, to collect these mountain plants the botanist must have a good nerve, a steady foot and eye, and must not be afraid of a little hard work and personal risk !

It is on these little damp rock ledges, often only a few square feet or even inches in area, where the moisture trickles down from the crags above, that we should look for the handsome and tolerably large Black Sedge (*C. atrata*). It grows on the mountains above Carnarvon, but is abundant on some of the Scottish Highlands, notably so on the Breadalbane and Clova mountains, which latter are perfect para-dises of Alpine flowers, conspicuous and inconspi-cuous, insect-fertilized and wind-fertilized. It is met with as high up as 3,700 feet. Whilst we are exploring these perpetually damp and somewhat protected rocky ledges, we may find the short Brown-spiked Sedge (*C. vaginata*). It is rather local in its distribution, and most abundant in the Breadalbane mountains. Nowhere is it to be searched for below the level of 2,000 feet above the sea. The Hare's Food Sedge (*C. lagopina*) grows in similar situations, but is so much rarer that the rambler may esteem it a botanical prize when he secures it. It flowers late in the summer, and may be easily recognised from its single long stalk and (usually) three tufts of flowers on the summit. Dr. Dickie first discovered it on the table top of Lochnagar, in the Aberdeenshire Alps, at a height of about 3,600 feet.

The drier parts of our Alpine rocks are not without their species of sedges, which appear as if they cared not for any other situations. Among these we may mention, first, the Rock Sedge (*C. rupestris*), which takes up its habitat on the dry rocky ledges in the mountains of Perth, Forfar, Aberdeen, and Sutherland, but seldom at a lower level than 2,000 feet. The stiff Mountain Sedge (*C. rigida*) is another lover of stony, Alpine localities, from North Wales northerly. It is found in the Highlands at elevations of more than 4,000 feet. The Vernal Sedge and Capillary Sedge (*C. præcox* and *C. capillaris*) also delight to occupy stony, rocky ledges, as well as the grassy hillside slopes. The moorlands and heaths have sedges peculiar to their often dry soils and short crisp pasturage. Among these we may mention *Eleocharis cæspitosa*, which may be found in such dry places as high up as 3,500 feet. The Round-headed

Sedge (*Carex pilulifera*) may be found abundantly growing in large dense tufts on our heaths; and its recognition is rendered easy by the characteristic way in which the slender flower-stalks grow until they bend over and their heads touch the ground. The Green-ribbed Sedge (*C. binervis*) is a spring plant also fond of flowering on our mountain heaths, where (in Scotland) we find it at elevations of more than 3,000 feet. The Scorched Alpine Sedge (*C. ustulata*) is perhaps the most rare of these numerous Alpine species, if indeed it be not now extinct in Britain. It was last found growing, in damp spots, on Ben Lawers, but frequent search for it there and elsewhere has failed to discover it.

The Mountain Rushes are not quite so numerous as the Sedges. Amongst the most notable of them, and those the observant tourist is likely to come across, are the Alpine Curved Rush (*Luzula arcuata*), so called on account of the manner with which the flower-heads droop, as if top-heavy. Its usual habitat is the stony *débris* on mountain-tops, at levels of not less than 3,000 feet. The localities where it is most abundant are the summits of the Cairngorm mountains. The Spiked Wood Rush (*L. spicata*) is a taller and far more abundant species, delighting in damp grassy situations at a much lower altitude, where it may be found in small tufts. The Welsh mountains and those of the Lake district yield this plant, but it is not so common in these localities as in Scotland.

The rocky ledges on the Scottish mountains, where we noted the presence of Alpine Sedges, should be examined for the Three-leaved Rush (*Juncus trifolius*). In the hilly bogs and moist places of the same elevations we may find both the Two-flowered and the Three-flowered Rushes (*J. biglumis* and *J. triglumis*), both of them very pretty plants. Both are rare, but the latter species is the rarer of the two. The much larger Clustered Rush (*J. castaneus*), another Alpine rarity, may be found in the same habitat. The Heath Rush (*J. squarrosus*), as its name implies, is peculiar to the drier parts. It is abundant on moorlands, ascending in the Highlands to as great heights as upwards of 3,000 feet.

The true grasses are the plants on which our British mountains depend for their greenness. It is they which carpet the hollows with verdure, and bring out into scenic relief the bolder features of crag and scaur. The mottled and variegated appearance of our hill and mountain sides is in this way produced by the grassed and ungrassed portions. And how soft and velvety is the mountain greensward! No couch seems half so comfortable to the tired limbs of the wanderer as these mountain resting-places.

We have just referred to the fact that the different parts of the flower of a grass correspond to the more showy parts which in a lily or a tulip arrest our immediate attention. The student can soon prove this for himself, by taking the flower of some common species, say the common Meadow Soft Grass (*Holcus lanata*), and carefully dissecting it. By doing so he will soon learn that the description of this species in a technical work on botany can be easily reduced to accurate simplicity. If the student only knows the difference between the floral parts, say of a buttercup, and can distinguish calyx from corolla, and the sepals or divisions of the former from the petals of the latter, he will have no difficulty in making out correctly the parts of a grass-flower.

In the *spikelet* of the Meadow Soft Grass, now under examination, the sepals and petals, although present, are altered so much in appearance that the student could not recognise them as the calyx and corolla, or what would be called by that name in flowering plants. Thus, on the outside, at the base, are two small, purplish, boat-shaped, bract-like organs, with **three** veins running up each : these correspond to **the** *calyx*, but are called *glumes ;* then next to these **are** the two *glumellas*, or *corolla*. The latter are **green,** therefore easily recognised ; they are also **much** smaller in size than the glumes, which in this grass are almost as transparent as a piece of glass, so that the glumellas can be seen through them. The term *perianth*, mentioned above, is applied to both the calyx and corolla, when it is difficult to distinguish one from the other; thus **the** glumes and glumellas are **the** protective organs of the flowers in grasses, or the perianth. Another name applied to the glumella in **some** botanical works is the *palea*.

In some **of** the spikelets may be detected both stamens and pistils ; in others only the stamens or pistils are present. The stigma, or the upper part of the pistil, is a very pretty, feathery appendage ; and the anthers, or the heads of the stamens, containing the pollen, or fertilizing dust, are very elegant, and are suspended on most delicate stalks.

A slight examination of the flowers of the upland grasses will convince us that the grasses of the hills and mountains are peculiar to such regions, and that they are quite distinct from those which produce such dense crops of hay in the warmer and more fertile meadows below. Some of these Alpine grasses are very scarce, and they rank among the prizes of botanical investigation. Others are so local in their distribution, sometimes confined to a single mountain-top or a part of it, that the student must make a long pilgrimage to procure them. Like the other Alpine plants we have referred to, these grasses are floral relics of the Glacial period.

Perhaps the most abundant and widest distribution of these upland grasses is that of the Mat-grass (*Nardus stricta*). On the moorlands it often constitutes the entire herbage for scores of acres together, particularly in the dry and heathy parts which it chiefly affects. In the Highlands of Scotland it grows luxuriantly as high up as 3,300 feet. The leaves are very fine and channelled, and the stiffish flower-spikes, when in flower, are very pretty, the clusters of yellow stamens being arranged on one side. It is in the herbage of this grass that the larvæ of the Mountain Ringlet butterfly love to feed. Another common species is the Wavy Mountain Hair-grass (*Aira flexuosa*), of which there are several well-marked varieties, adapted to as many different physical conditions. The flower-stem is very graceful, owing to the slenderness of the wavy, hair-like branches of the panicle. The Alpine

Hair-grass (*Aira alpina*) is neither so graceful in shape nor rich in the colour of the flower-panicle. It is a rare species, frequenting the rocky *débris* of mountain summits, and most abundant in the Braemar mountains and on others in the Highlands.

The well-known Sheep's Fescue-grass (*Festuca ovina*) is an abundant species on our heaths and moorlands. Cows, horses, and goats will eat it, but not from preference. With sheep, however, it is the favourite food. They prefer it to all other grasses, and are said soonest to grow fat upon it. There are a great many distinct varieties of this common mountain grass. One of these is remarkable for its habit of not shedding the ripened seeds. On the contrary, they remain where they were developed, and sprout there, dropping off to the ground and taking root immediately. This method of reproduction is called *viviparous*, and although it is so frequently adopted by the higher mountain variety of the Sheep's Fescue-grass as to have given it this name, it is not limited to this species, but is a plan resorted to by the varieties of several other Alpine grasses.

Some of our common meadow grasses (*Poa*) ascend to great heights up the hilly slopes, notably *Poa annua* and *P. pratensis*. But this genus includes several rare species, to be found only in odd localities on rocky ledges, or among the stony *débris* near the summits of our highest hills and mountains ; such as *Poa laxa* (met with on the lofty Alps of Aberdeen and Inverness) and *Poa alpina*, which is more abundantly distributed, for we may gather it on the tops of nearly all our English, Welsh, and Scotch mountains.

The specific name of *Alpina* at once indicates the grasses and other plants which affect these elevated and bleak localities. Thus we have the Alpine Foxtail (*Alopecurus alpinus*), which grows on the margins of springs and streams on the mountains of Ross, Aberdeen, Perth, Forfar, and Inverness; and the Alpine Timothy-grass (*Phlæum alpinum*), which also grows near the edges of mountain springs and rills on the Alps of Forfar, Perth, and Aberdeen; a well-marked, and very distinct variety of the Oat-grass (*Avena pratensis*), is by some well-known botanists elevated into a species under the name of *Alpina*. It grows on most of our higher British hills, moorlands, and mountains.

Of the rarer and extremely local mountain grasses, much sought after but seldom found, are *Phlæum Michelii* and *Poa Balfourii*. Indeed, the former grass is said to have been only found by one man, Mr. G. Don, growing on the summit of the highest mountain in Forfarshire. The second grass is named after the veteran professor of botany in the University of Edinburgh, than whom no one in Great Britain is better acquainted with the Alpine flora of Scotland. It is found on rocky *débris* and ledges on the Snowdonian range, in North Wales; on Ingleborough, in the West Riding; on the summits of the Cheviot Hills, in Northumberland; and on many of the Scottish Alps.

On limestone hills we may probably find the Blue Moor-grass (*Sesleria cærulea*), so called from the bluish-grey colour of the small and dense panicle. When it is in bloom, and the yellow stamens are dangling outside the blue glumes, it is really a very pretty grass. It is found on the Carboniferous limestone hills of Lancashire and the West Riding of Yorkshire, as well as on those of Westmoreland and northwards into Scotland.

The worst and poorest of our heathy and moorland soils are not unrepresented by grasses. There we are almost sure to find the Decumbent Heath-grass (*Triodia decumbens*), particularly in the driest and most arid spots. It is itself a poor, poverty-stricken looking plant as regards its flower-stem, but this is atoned for by the bright green of its leaves. By way of comparison, we conclude with drawing attention to the exceedingly graceful flower-spikelet of the Nodding Melic-grass (*Melica nutans*), which must be sought for in the woods and shady banks at no great height up the hill or moorland side.

Enough, however, has been said, even of these inconspicuous, uncoloured, and little perfumed kinds of flowers, to indicate to the reader that, even as regards their grassy carpeting, there is provided for "the cattle upon a thousand hills" a special provender, seldom or never met with in the richer pastures below, and having an origin as different from that of the latter as the Caucasian differs from the Semitic race of mankind. Science demonstrates the truth of the Wise Man's proverb, that if "It is the wisdom of God to hide a thing, it is the glory of a king to find it out!"

CHAPTER VIII.

THE FERNS, MOSSES, LICHENS, ETC., OF OUR HILLS AND MOUNTAINS.

Fern-hunting—The Bracken—Its World-wide Distribution and
Geological Antiquity—Hard Fern or *Lomaria*—Wall Poly-
pody—Beech Fern—Oak Fern—Alpine Polypody—Alpine,
Mountain, and Common Bladder Ferns—Wood's Ferns—
Filmy Ferns—Holly Ferns—Shield Ferns—Parsley Fern—
Spleenworts—Male Fern—Mountain Fern — Stiff Fern —
Moonwort—Horse-tails—Quill-worts ; Geological Antiquity
of Club-mosses—*Selaginella*—Superstitions regarding Club-
mosses—Liverworts—Jungermannias, or "Yoke-mosses"—
Their habits and structure—Bog-mosses or *Sphagnums*
—Their Functions in Mountain Geography—Alpine Mosses
—Lichens—Their Variety and Geographical Distribution—
Confervæ, Diatoms, &c.

Of all the holiday engagements which have a flavour
of science about them, none is so popular as fern-
hunting. More than one resort of tourists owes its
thriving condition to the ferns which grow in the
neighbourhood. To see the numbers of people
returning from their summer holiday in the Lake
district, with hampers and baskets crowded with
living plants of various species of fern, which they
are carrying home, is enough to prove that ferns possess an attraction beyond that of any other class of
plants. Nor do we wonder, for none other can
compete with them in their fresh greenness, elegant
shapes, and graceful mode of growth.

The exceeding dampness of mountain districts is greatly in favour of the growth of ferns. Everybody knows how the common Bracken (*Pteris aquilina*) mantles the hill-sides with its tender green in June, changing it to a duller tint in July and August, and breaking out into a mass of yellow, orange, and red in October and November. Common though this fern be, it is one of the most difficult to cultivate artificially. It will sprout up amid the entanglement of thick hedges, dense branches of gorse, sterile stony heaps, sandy slopes that will grow nothing else, bleak exposed moorlands, and as high up the mountain slopes, even in the Highlands of Scotland, as 2,000 feet; but it can with the greatest difficulty be coaxed into growing in Wardian cases, in rich soils, under careful and anxious human protection!

Mr. H. N. Moseley, F.R.S., in his recently-published work, "Notes of a Naturalist on the 'Challenger,'" expresses his surprise at the world-wide geographical distribution of the Bracken fern. It is found growing, almost equally abundantly and evidently quite "at home" in arctic, temperate, and tropical climates. Perhaps this cosmopolitan distribution is indicative of its geological antiquity. One genus of abundant fossil ferns found in the coal-measures (*Alethopteris*) resembles the modern Bracken in almost every particular, and may possibly be its lineal ancestor.

We love this dear, honest, old Bracken, which, in its free, unfettered and unfetterable mode of growth, seems so well suited to the liberty of the heaths and moorlands! Nor is this abundant fern without

its use, if people only knew it, for when cut in the
autumn, dead and dried, and employed to roof over
potato-heaps instead of straw, it will invariably protect them from rot and disease ! Every one is aware
that if the stem of the Bracken be cut across, the
woody bundles of vessels in them look like the heraldic
figure of a spread eagle. It is in reference to this
that the plant has obtained its Latin specific name of
aquilina.

Nearly allied to the Bracken is the Hard Fern
(*Lomaria spicant*). It is, however, distinguished from
it by the central fronds, which we term " fertile,"
because they alone bear the spore-cases. They
are crowded with them, and hence are stiff and
erect ; the pinnules also being much narrower than
those of the " barren " fronds, which occupy the outer
part of the whole plant, and usually lie on the ground.
These outside fronds are usually of a dark, glossy
green ; and the outline of each is elegantly lancet-
shaped. The Hard Fern is more commonly distributed to the north than in the south, and delights
in heathy or moorland situations, although it never
grows in such immense numbers as the Bracken.

The Polypodies, although some of them are abundant in the green lanes and woods of the lowlands,
may also be regarded as hill and mountain ferns.
The common Wall Polypody (*Polypodium vulgare*)
is found in the West Riding of Yorkshire at elevations
of more than 3,000 feet. The less abundant, and
more decidedly mountain species of this genus are
the Beech Fern (*Polypodium phegopteris*), which should
be looked for in the damp, shady places along the

hill-sides, in and out of whose moss-covered stones its wiry rootstock will be found creeping. It is very abundant in North Wales and the Lake district. The Oak Fern (*Polypodium dryopteris*) is, in our opinion, a much more lovely species, for its light green fronds are of an elegant triangular shape, and the lovely tint of the greenness seems to come out all the better by reason of the slender, shiny, almost black stipes or stems. **The damper parts of** the shady mountain spots are the places **to look** for it, especially where the ground is encumbered with detached fragments or lichen-covered boulders, for at their bases and **under** their shelter it will be seen growing. The Beech Fern grows as high up in the Highlands as 3,500 feet; but the Oak Fern is not so hardy, and is seldom met with at a greater height than 1,000 feet below where the Beech Fern finds it possible to flourish. The Alpine Polypody (*P. alpestre*), as its name imports, is only to be met with by severe mountain-climbing, **and** at the cost of much labour and careful examination. It is limited to the Scottish mountains, and usually occupies the zone of from 1,200 to nearly 4,000 feet. The fronds rise to the height of 6 inches, each **pinnule** is delicately cut or serrated, and the general appearance resembles a dwarfed specimen of the graceful Lady Fern.

Whilst we are examining the moist clifts and crannies of Alpine crags, we may as well look out for other species of ferns we certainly shall not find at a much lower level, or under other than Alpine conditions. Among these are the Alpine and Mountain Bladder Ferns (*Cystopteris alpina* and *Cystopteris*

montana). Both grow in delicate tufts, of from four
to six or seven inches in length. The former is exceedingly rare, only one or two British localities
having been as yet recorded for it. The latter is
more abundant by comparison, but still rare enough
to be esteemed a "good find," and only to be met
with on the Scotch mountains, notably on those of
Clova and Breadalbane. Both are exceedingly deli-
cate-fronded ferns, and the wonder is that such fragile
things should prefer to grow only where the situation so nearly resembles that of the Arctic circle!
The Brittle Bladder Fern (*Cystopteris fragilis*) is
scarcely less Alpine in its hardiness and geographical
distribution, for we find it growing at the height of
4,000 feet in the Highlands. It appears to prefer limestone regions, and in the Peak of Derbyshire it is one
of the very commonest of ferns. Its popular name is
in allusion to the brittle character of the stipes or
stems—a fact the fern-grower soon finds out when he
attempts to remove this graceful fern. It has a rather
extensive distribution in all the mountainous districts
of Great Britain. Wood's Alpine Fern (*Woodsia
hyperborea*) is another delicate and diminutive mountain species, growing in tufts of two or three inches
in height, in the clefts of the high and almost
inaccessible rocks of the mountain summits, in
the neighbourhood of Carnarvon, North Wales, as
well as in similar situations in Scotland. Wood's
Hairy Fern (*Woodsia ilvensis*) is also an Alpine
species, to be searched for on wet rocks, anywhere
on high mountains from North Wales and Durham
to Dumfries, Perth, and Forfar. It grows to about

the same size as the preceding species, but may be distinguished from it by the bristly hairs on the under surface of the fronds. In the shaded part of these damp rock-crevices we shall probably find Wilson's Filmy Fern (*Hymenophyllum Wilsoni*, or *H. unilaterale*, as it is also called). It is a most delicate and fragile-looking plant, more likely to be taken for a moss than a fern ; and, if found in fruit, an examination of the vase-shaped sporangia or spore-vessels will

WILSON'S FILMY FERN (*Hymenophyllum Wilsoni*).

be worth the trouble. This fern is found on the Irish, as well as the English, Welsh, and Scotch mountains, but always at high elevations.

In the drier cliffs of these lofty crags, and amongst the *débris* of large stones, we shall find the Holly Fern (*Aspidium lonchitis*), whose length of fronds will vary, according to the suitableness of its habitat, from 6 to 18 inches. Their *prickly* appearance, however, will be a good guide to their recognition, for in this case, for once, popular nomenclature has been very happy. The fronds are of a dark bright green colour; the pinnules are packed so close that they partially overlap one another. Most of our British mountains grow this characteristic Alpine species of fern. The Prickly Shield Fern (*Aspidium aculea-tum*) also grows high up the damp and shaded parts of the hills, but it never attains Alpine altitudes. It is altogether a much commoner species, and the large number of glossy pinnules arranged on the frond are much smaller in size than those of the true Holly Fern.

The Parsley Fern (*Allosorus crispus*) is quite a mountain species. It does not hide in damp, dark, nooks and crannies of rocks, however, but delights in the open mountain-side, seeking only such shelter along the slopes as may be obtained beside the huge boulders which have been strewn about. The Parsley Fern has two sorts of fronds, one barren and the other fertile. The fronds appear first in the month of May, and nothing can then exceed the fresh greenness of this lovely fern. These parsley-like barren, or "vegetative" fronds are developed before the "reproductive," or spore-bearing fronds. Many of the mountains in the Lake district are greened over by the luxuriant growth of the Parsley Fern, every stone and boulder being set in a framework of its surrounding greenness. It is also to be met with, less abundantly, in odd places on the Welsh mountains, on Dartmoor, the tops of the Pennine chain of hills, in the Highlands of Scotland, and elsewhere.

The Spleenworts are a very remarkable, rock-loving group of ferns, some species of which are sure to be

M

found in any of our British mountainous regions.
The commonest of these is the Maiden-hair Spleenwort (*Asplenium trichomanes*), to be found in
abundance in the crevices of the stone walls and
rocky crags in most hilly districts, at as great heights

THE GREEN SPLEENWORT (*Asplenium viride*).

as 2,000 feet above the sea-level. Its black stem
distinguishes it from a much rarer allied species,
which has a slender *green* stem, and on that account
is named the Green Spleenwort (*Asplenium viride*).
It appears to love limestone hills better than those
of any other kind of rock, at least such is our expe-

rience of its occurrence.   In the Highlands it grows
up to altitudes of nearly 3,000 feet.   The Rue-leaved
Spleenwort (*Asplenium ruta-muraria*) has a very
extensive distribution, high and low, and is one of
the best known of our small ferns, to be found in the
mortar-crevices of most old churches and ruins, and
in those of dry rocks in hilly situations.   The Black

FORKED SPLEENWORT (*Asplenium septentrionale*).

Maiden-hair Spleenwort (*A. adiantum-nigrum*) is
another fern with a black stem, much thicker than
the rest, however, and it will be found flourishing
beneath rocky ledges up to the height of 2,000 feet.
It is a common, although exceedingly handsome
and cheerful-looking fern.   Two much rarer species,
nearly allied to the Rue-leaved Fern, are the Forked

Spleenwort (*A. septentrionale*) and the Alternate
Spleenwort (*A. Germanicum*).

The former fern has, we believe, only been found
in the North of England, and on or near Cader
Idris, North Wales, in which district also the latter
has been met with, for it has been noted, singularly
enough, that these two species usually occur together.

ALTERNATE SPLEENWORT (*Asplenium Germanicum*).

But the Alternate Spleenwort has also been gathered
in North Devon, Somerset, North Wales (at Llan-
rwst, and in the Pass of Llanberis), and in the Lake
district. Some of the Male Ferns (formerly called
Lastrea, and now known as *Nephrodium*) are hillside
or moorland frequenters, particularly *Nephrodium
oreopteris*, popularly known as the 'Mountain
Fern," from its being found in such abundance on

elevated moorlands and elevated slopes. Its hay-scented perfume when trodden upon renders its identification easy ; or it could be drawn through the hand in order to elicit its fragrant odour. If this be not sufficient to enable the collector to recognise it, he should notice its erect, shuttlecock-like habit of growth. It is an abundant fern on the Lancashire and Yorkshire hills, as well as on the Welsh mountains, and in the Highlands of Scotland it ascends to nearly 3,000 feet. In Ireland, however, it is very rare, and confined to a few localities. The well-known and widely distributed male fern (*Nephrodium Filix-mas*)—the plant which in all old books is referred to as *the* fern—may also be said to be a mountain species, as it also is an abundant one in the plains, for in Yorkshire it grows as high up among the hills as between 2,000 and 3,000 feet. Another species, the Stiff Fern (*Nephrodium rigidum*) is much rarer, and more characteristically an upland form. It grows on the elevated rocks and in the mountain districts generally of Lancashire, Yorkshire, and Westmoreland, especially on the Carboniferous limestone region, which is continuous in those three counties.

Another plant which, if not actually a fern, must be classed among that family, is the once well-known Moonwort (*Botrychium lunare*). The botanical name is a Greek word signifying a " little bunch of grapes," and is an allusion to the clusters of spore-cases borne upon the spike, much after the manner in which grapes are grown. The true fern it is most allied to is the Royal Flowering Fern, which, it will be remem-

bered, bears its brown spore-cases on separate upright
branches on a larger scale, in a similar manner. The
Moonwort is perhaps the best known of all the fern
family. Chaucer alludes to it as part of the alchemist's
furniture,—

> " And herbes coude I tell eke many on,
>   As Egremonie, Valerian, and Lunarie."

The word "lunarie," however, may refer to a very
old English flower, common in cottage gardens under
the name of "Honesty," and which was also formerly
known as "Moonwort." The Ettrick Shepherd certainly alludes to our "Moonwort" :—

> " We saddled our naigis wi' the Moon-fern leif,
>   And rode fra' Kilmenin Kirk."

The old idea concerning it was that people possessed
of any sort of "fern-seed" could render themselves
invisible.

This "seed" was the *spores* of ferns, which all
botanists are aware are very minute—hence, perhaps,
the idea as to invisibility. But it was believed that
these microscopic spores could be seen on St. John's
Night, at the hour when the Baptist was born ; and
hence, whoever became possessed of them was thereby
rendered invisible. This idea is given us in the
following lines :—

> " But on St. John's mysterious night,
>   Sacred to many a wizard spell,
> The time when first to human sight
>   Confest the mystic fern-seed fell :
> Beside the sloe's black knotted thorn
> What hour the Baptist stem was born—
> That hour when heaven's breath is still—
> I'll seek the shaggy fern-clad hill,

Where time has delved a dreary dell,
Befitting best a hermit's cell ;
And watch, 'mid murmurs muttering stern,
The seed departing from the fern,
Ere wakeful demons can convey
The wonder-working charm away,
And tempt the blows from arm unseen,
Should thoughts unholy intervene."

Brand states in his "Popular Antiquities" that "a respectable countryman, at Heston, in Middlesex, informed him in June, 1793, when he was a young man, he was often present at the ceremony of catching the fern-seed at midnight on the eve of St. John Baptist. The attempt, he said, was often unsuccessful, for the seed was to fall on the plate of its own accord, and that too without shaking the plant."

The *Botrychium* should be looked for on the grassy pastures of the hills, where it loves most to grow. In the Highlands it is found as high up as nearly 3,000 feet.

The margins of mountain-tarns, at high elevations, should be searched for a peculiar species of **Horsetail** (*Equisetum limosum*), which, in Great Britain, is peculiar to elevated regions. Another inhabitant of the same lakes, and of others at even greater altitudes than this kind of Horsetail can grow in, is a species of Quillwort (*Isoetes lacustris*). It grows on the *bottoms* of these mountain pools, and must be dragged for. When brought up, notice the bulb at the base of each grassy, awl-like frond, how it contains a host of white spore-cases. It is also found at the bottoms of all our British Alpine and sub-Alpine lakes, from North Wales into Scotland, and is met with in Irish lakes, where it is believed the trout feed on the tender young shoots.

A class of cryptogamous plants, which is unquestionably mountainous in its habits, is that of our British Club Mosses (*Lycopodium*). They are the dwarfed representatives of the oldest plants on the globe, for some of them made their appearance in the geological period known as the Silurian. They are therefore probably the oldest of terrestrial plants. During the Carboniferous epoch, these plants grew to the height of forest trees, some of them as high as 80 feet. With us in Britain, club mosses are essentially mountain plants, and appear to delight in elevated situations. This may be because they find there fewer competitors, and a better means of dispersing their spores, on account of the breeziness of the mountains and moorlands. The Stag's-horn Club Moss (*L. clavatum*) is one of the largest, and most abundant, its rope-like stems sometimes trailing on the ground for a great distance, and actually setting a trap for the feet of the incautious sportsman or rambler. It affects the drier parts of the hills, but is seldom found higher up than 2,500 feet. The species which grows at the greatest height is the Alpine Club Moss (*L. alpinum*), which is not uncommon on the open, stony moors, heaths, and other dry, elevated places in North Wales, Derbyshire, Lancashire, Yorkshire, Westmoreland, and Scotland. In the Highlands it may be found growing at 4,000 feet above the level of the sea. *L. annotinum* is also fond of elevated stony moors, and may be found growing in these places almost anywhere north of

Leicester.   It is a smaller, but prettier species than
the Stag's-horn Club Moss.   *L. selago*, another mountain club moss, will be met with on nearly all the hill-
tops of Great Britain, from Sussex and Cornwall to
the Highlands of Scotland, on the crags of which

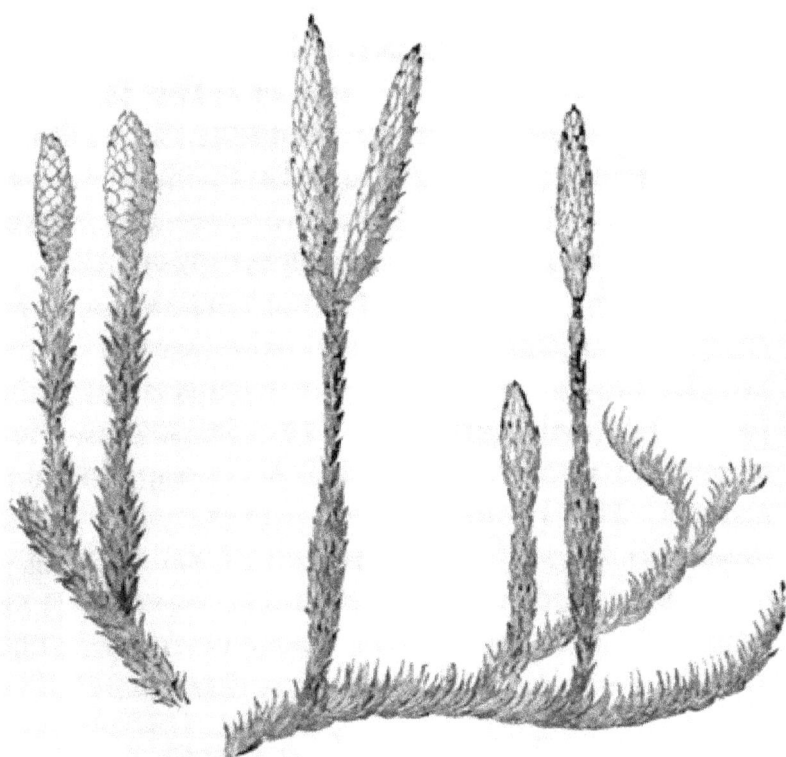

STAG'S-HORN CLUB MOSS (*Lycopodium clavatum*).

latter it is found growing as high as 3,500 feet.   It
may easily be known from the other species by its
stronger and thicker branches.   *L. inundatum* differs
from all our other British species by its habitat.   It
prefers wet heaths and bogs to the stony and dry

places, and its entire appearance is therefore much greener than any of the rest. One only British species of *Selaginella*—a plant nearly allied to the club mosses, may also be found in the bogs and marshes of northern England and Scotland; in the latter country at very great altitudes. Its colour is of a tender, yellowish-green tint.

Club mosses were formerly deemed good for all complaints or diseases of the eyes. The Bog Club Moss (*L. inundatum*) occurs plentifully in Cornwall; and Mr. Hunt tells us that the gathering of this species there was conducted with great secrecy and mysteriousness. So important was this method, that we can credit the plant with very little actual virtue, inasmuch as all of the latter seemed to be absorbed in the peculiar manner with which the moss was gathered. It was believed that if any one were to write the secret, the virtues of the Club Moss would all disappear! Mr. Hunt, however, has not scrupled to reveal this extraordinary secret, which was as follows: "On the third day of the moon, when the thin crescent is seen for the first time, show it the knife with which the moss is to be cut, and repeat—

" As Christ healed the issue of blood,
    Do thou cut what thou cuttest for good.

At sun-down, having carefully washed the hands, the Club Moss is to be cut kneeling. It is then to be carefully wrapped in a white cloth, and subsequently boiled in some water taken from the spring nearest to its place of growth. This may be used as a fomentation. Or the Club Moss may be made into an

ointment with the butter made from the milk of a
new cow !"

In the damp places among the rocks, on the stones
of the rill-sides, and in a thousand unnoticed places
by their sloppy banks,
we may find in the
summer-time a host of
species of Liverworts
and *Jungermannias*.
Their bright-green, al-
most emerald tints, light
up the dim hollows, and
convert them into minia-
ture fairy grottos. The
huge stones in the beds
of the stream are uphol-
stered along their lower
sides with *Jungerman-*

COMMON LIVERWORT (*Mar-
chantia polymorpha*).

nias.   Notice the simple yet lovely cellular tissues
of their fronds. Those of the latter order of plants
are almost transparent. Both kinds have peculiar
modes of reproduction, and in the late summer the
observer will find them elevating their star-shaped,
mushroom-like spore-vessels on their slender green
stalks. We have several species of these Liverworts,
all of them peculiar to wet places, and each is distinguished by the shape of its spore-cases. All of them
alike possess dark, emerald-green, lichen-shaped
fronds, or leaves, often beautifully marked with
polygonal cells. The Hemispherical Liverwort (*Marchantia hemisphærica*) is a plant of denser and
more luxuriant growth, distinguished from the

former by this, and the shape of the mushroom-like capsules.

The "Yoke Mosses," or *Jungermannias*, are exquisite objects, almost limited to mountain streams and rills and constantly-dripping rocks. It is only of late years that botanical attention has been directed to them. They are delicate-looking objects, owing to their loose, cellular tissue ; and this latter fact causes them to fade and wither very rapidly after being

THE HEMISPHERICAL LIVERWORT (*Marchantia hemisphærica*).

gathered. Some of them have a wonderful fern-like appearance, particularly simulating the appearance of the Film Ferns (*Hymenophyllum*) ; but the botanist is well aware that their structure and habits are altogether different, and that this is another illustration of the adage, that appearances are often deceptive. Our British species furnish us with examples in which it is not difficult to trace the gradual transition from the liverwort-shaped frond to that almost resembling the film ferns, as in the case of *Jungermannia compla-nata* and *J. inflata.*   The size of the different species varies considerably, one of the largest being *J. pinguis,* whilst others are so diminutive that they can only be

JUNGERMANNIA PINGUIS (nat. size).

found by the keenest search.   Most of the species it is useless to look for except in Alpine situations.

The true mosses, as every one is aware, also delight in mountain habitats, where the constant moisture which gathers around the flanks and summits is a great promoter of their growth. Hence a very large number of species of our mosses can only be found in the nooks and crannies of the mountains. One group of mosses, the *Sphagnums*, has become specialized to marshy conditions, and so these are popularly termed "Bog Mosses." Everybody is acquainted with their yellowish-green, woolly appearance, often masking the most deceptive and treacherous of the boggy places on the hills and moorlands. The bright, shining, black fruits are borne on stalks on the surfaces of the feathery bog mosses in the summer time, although these plants are not dependent on fructification for their propagation, but spread themselves by the simpler habit of "budding." This latter habit is also much better suited to the semi-aquatic conditions in which they are found.

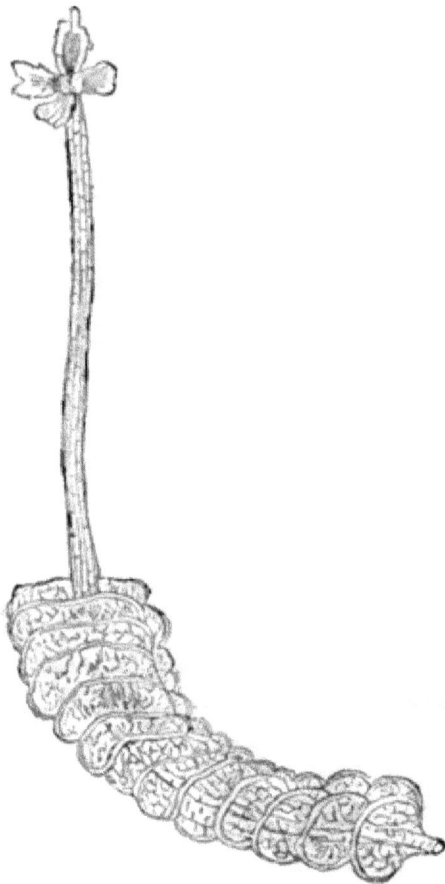

JUNGERMANNIA COMPLANATA.

The inhabitants of the plains and lowlands owe

more to these Bog Mosses than they are aware of, for they are important factors in a very practical and beneficial physical geographical operation. As every one who has been among the mountains is aware, the lower parts and bases are usually occupied by thick and extensive sloppy bogs and marshes. It is here that the *Sphagnums* will be found mustering in the greatest force, and where they are of the greatest service. The heavy rains falling on the hard and almost impenetrable rocks, flow down the hillsides and mountain-flanks, and, gathering force as they descend, would rush into the plains below and commit much havoc and devastation, were it not for the bog mosses through which the descending water has to percolate. These *stay* its progress, and delay it so that only an enfeebled and diminished volume is always oozing or flowing from the lower end of such mountain marsh. These mosses are therefore great regulators of the rainfall of hilly districts, as well as ameliorators of their physical geography. And in the heat of summer, when the lowlands are scorched and parched for want of rain, there will always be found cool clear rills trickling forth from these mosses, to supply the streams which descend to the plains below.

Our Alpine mosses are very diminutive and of local distribution. The search for and discovery of them are among the most harmless and enjoyable of the botanist's summer wanderings. Some mountain summits are notable for the number of Alpine mosses which grow on them. The Highlands of Scotland stand first in every respect, Ben Lawers (and others of the mountains of the Breadalbane range), Ben Nevis, Ben Cruachan, the Clova mountains, Braemar, Ingleborough Fell, Teesdale, Wharfdale, the Peak of Derbyshire, and the mountains of North Wales. Some genera of British mosses contain more truly Alpine and sub-Alpine species than others; and amongst the most noteworthy in this respect are *Bryum*, *Hypnum*, *Orthotrichum*, &c., whilst the Hair Moss (*Polytrichum*) is more characteristic of our moors and hilly copses. Not a few species are limited to one or two solitary habitats, as *Andræa alpina* on Ben Nevis, and *A. Rothii* on the same mountain and the Cairngorms. Here and there an odd species, unknown in few other spots, will be found on the dung-heaps of foxes or deer. At least *twenty* genera of British mosses are represented by Alpine species, and some of these by several distinct kinds. Each genus may be distinguished by its peculiar kind of fruit, or *sporangium*—the equivalent of the spore-cases on the backs of ferns, and of the bulbous swelling of the bases of the green awl-like fronds of the "Quillworts." When the preliminary difficulties of recognition are surmounted, few outdoor studies are more fascinating than that of muscology. The student never need be without objects, wherever he may go; although it is in the solitude of the hills that he finds his richest harvest of unbloody spoils.

There is much philosophical as well as religious truth in the following lines, and the application need not be confined only to the objects which suggest it :—

" The tiny moss, whose silken verdure clothes
The time-worn rock, and whose bright capsules rise,
Like fairy urns, on stalks of golden sheen,
Demand our admiration and our praise,
As much as Cedar kissing the blue sky,
Or Krubert's giant flower. God made them all,
And what He deigns to make should ne'er be deemed
Unworthy of our study and our love."

But if our mosses abound most in mountain
districts, what shall we say of the Lichens? Everybody knows how they not unfrequently mat the ground
with their blackish-grey fronds. They upholster the
hardest rocks with their rich mantle of brilliant yellow,
orange, and silver-grey. Their variegated patches
relieve many an otherwise bald and unpicturesque
crag and boulder, and thus convert objects that
would otherwise be disagreeable to the sight into
" bits " which artists come from far and near to study,
and transfer to " foregrounds." Even the white ribs
of quartz which seam the mountain-peaks, unim-
pressible almost by the storms which spend their
fury around, find a foothold for crustaceous lichens,
and fall a prey to their slow but certain power of
decomposition. Jane Taylor but expresses the un-
uttered opinion of every lover of the mountains who
has observed how these humble and lowly members
of the vegetable kingdom throw a mantle of beauty
around them :—

" Art's finest pencil could but rudely mock
The rich grey lichens broider'd on a rock."

The wandering botanist has the enchanter's power
of discovering a garden where others see only a

N

wilderness; instead of bleak desolation spreading over our mountain-peaks and stretches of moorland, he finds there living objects, sharing the Creator's bounty, and indicating in their structures His ineffable wisdom, of which the world little dreams ! All around him, thus high above the busy plains below, where the fierce fight for wealth and power is everlastingly going on, there is a veritable "Garden of the Lord," crowded with *Lecanoras*, *Lecideas*, *Umbilicarias*, *Cladonias*, and many another genus of lichens he knows would never be found two or three thousand feet beneath ! On the top of Ben Nevis, Ben Lawers, or Ben Lomond, he searches for and finds the beautiful *Solorina crocea*, whose dull green surface is more than compensated for by the rich saffron-colour of the underside, which has earned for this lichen its specific name. Many of the lichens have their local distribution determined by the nature of the rocks. Thus, the species just mentioned seem to be fond of granite ; others are equally peculiar to slate or lime-stone. No other group of plants, except perhaps sea-weeds, have such a world-wide geographical distribution as the lichens, although they are everywhere characteristic of zones, or of latitudes, where cold and abundant moisture are present. Out of 300 species gathered by Sir James Ross in the Antarctic regions, more than one-half were found to be identical with those growing in Europe. Recently a new interest has been given to lichens through Schwendener's theory, that the spore-cases contained within the substance of the fronds are in reality imprisoned algæ.

A good deal of controversy is taking place over

this " Lichen-gonidia " theory just now. Schwendener
contends, not only that all lichens are algæ, which
have collected about them parasitic fungi, but that
the *gonidia* or spore-cases are only imprisoned algæ.
His statement is as follows :—" As the result of my
researches, all these growths are not simple plants,
not individuals in the usual sense of the term ; they
are rather colonies, which consist of hundreds and
thousands of individuals, of which, however, only
one acts as master, while the others, in perpetual
captivity, provide nourishment for themselves and
their master. This master is a fungus of the order
Ascomycetes, a parasite which is accustomed to live
upon the work of others ; its slaves are green algæ,
which it has sought out, or indeed caught hold of,
and forced into its service. It surrounds them, as a
spider does its prey, with a fibrous net of narrow
meshes, which is gradually converted into an impenetrable covering. While, however, the spider
sucks its prey and leaves it lying dead, the fungus
incites the algæ taken into its net to more rapid
activity ; nay, to more vigorous increase."

The limits of space prevent us doing more than
drawing attention to a fact not sufficiently understood, viz., that the highest peaks of our mountains, the bleakest sides of our hills, and the most
desolate-looking of our moorlands and upland heaths,
have vegetable inhabitants of their own, adapted to
every physical condition of Alpine and sub-Alpine
existence. The most highly organized of these
possess, as we have already seen, a geological as well
as a botanical history of their own. The most lowly

organized, even of species of plants, we have not been able to notice, such as the *confervæ* or green " silk-weeds," which mantle the surface of every stagnant tarn ; or the lovely *diatoms* which may be wrung out of a handful of damp moss, and found in abundance in every pool, but whose minute sculpturing on their glassy shells demands the highest microscopical power which man can bring to bear upon them,— all these find representatives on Alpine heights or elevated plateaus. Not alone do they live there, as fern and lichen and moss, to cheer the eyes of those occasional wanderers who take note of them ; they have a work of their own to do in the economy of the physical universe, and the poet's view is both scientific and philosophical, when he declares of such humble and inconspicuous objects, that each

> " Holds a rank
> Important in the plan of Him who framed
> This scale of beings ; holds a rank which, lost,
> Would break the chain and leave behind a gap
> Which Nature's self would rue ! "

CHAPTER IX.

THE MAMMALS OF THE MOUNTAINS AND MOORS.

Relations between Alpine Plants and Animals—The Red Deer—
Its former abundance—Its stages of development—Professor
Bell on Life-history of Red Deer—The Roebuck—The Alpine
Hare—Manner of changing its colour of fur—The Wild Cat
—Difference between wild and domestic cats—The Pine
Marten and Beech Marten—Common Marten—Food and
Habits of the Marten——The Ermine Weasel—Pole-cat—
The Otter, Badger, Rabbit, &c.—The Irish Hare—The Fox
—Probable extinction of Wild Cat and Pine Marten—Ex-
tinction of Bear and Wolf in the Highlands.

A FEW mammals, and a larger and more varied
assortment of birds, as well as insects of several
orders, find in the solitudes of our hills the protection
and immunity they are denied in the plains. Many
of the insects are attracted here, and induced to
make these elevated regions their permanent resting-
place because of the Alpine and sub-Alpine flowers
and grasses. Some of the birds follow these insects
and prey upon them, whilst not a few kinds live
wholly on the seeds and berries of upland plants.

Our mountain mammals are now very few in
number, except where they are protected for purposes of "sport." There can be little doubt they
were formerly much more abundant, for many kinds
have been all but decimated, in order to render that

"sport" more effective.   Man denies the right to kill to any other animal.   None can do it so effectively, nor on such an extensive scale, as himself !

A reference to the mammalia inhabiting our mountain strongholds unconsciously brings up to our minds the Red Deer of the Highlands (*Cervus elaphus*). Time was when this noble animal was not confined to the few habitats or "deer forests" where, by the grace of mankind, it still runs wild.   The references to it in those ancient ballads which chronicled the doings of Robin Hood and his merry men, as well as the existence of the "Forest Laws," shows how much more abundant it must have been in the wild state a few hundreds of years ago.   The number of antlers and skulls of the Red Deer found in peat formations also plainly prove their pre-historic abundance and former extensive distribution.   In the Middle Ages, the "noble science of venerie" made the deer a most important animal.   Terms even more rigid than those employed in modern scientific nomenclature were invented, to betoken the different ages and stages of both sexes of deer.   The young were all termed *calves*.   In the first stage of development the young male or stag was called a "knobber"; in the second year it had been advanced to a "brocket"; in the third year it had become a "spayad"; in the fourth a "staggard"; and it was not until it had reached its fifth year that it was correct to term it a "stag."   Even then the hunter's nomenclature had not ceased, for afterwards, when the stag developed its hind antler, it had arrived at the dignity of "royal"; when it had acquired its sixth antler it was a "sur-royal." Similarly, the female deer, from its young state up to its adult, was known by various terms, intended to be expressive of the several changes. Prof. Bell, in his "History of British Quadrupeds," gives the following summary of the life-history of the Red Deer :—" The pairing sea-

RED DEER—HIND AND FAWN.

son is in August, and it continues about three weeks. During this period the harts are in a state of extreme excitement, and fight furiously when two of the same age and similar strength happen to meet. The hind goes with young eight months and a few days, and seldom produces more than one calf. The hinds then retire from the herd to bring forth, and continue

to attend their young with the greatest assiduity and
tenderness ; these remain with their dams during the
summer, and in the winter the whole herd becomes
completely gregarious. About February the males

RED DEER (*Cervus elaphus*).

lose their horns, and they begin to be renewed
shortly afterwards. At this period, and for some
time subsequently, they retire from the herd and
remain apart."

As we have said, the only places where the British

Red Deer is still in a wild state is in the Highlands of Scotland. The mountains which rise from the elevated wild plateau of Rannock Moor, up to which the celebrated Pass of Glencoe leads, are their favourite haunts. This is the late Sir Edwin Landseer's "sketching country," and most of the wild scenes, in which he portrayed the Red Deer, were studied in this neighbourhood. Black Mountain is, perhaps, the best part of this "deer forest," which belongs to the Marquis of Breadalbane. We have seen the deer, male and female, come out at even-tide, just in the gloaming, and have beheld a herd standing out on the top of the hills, clear and sharply cut against the grey sky. In the late summer and autumn, when the bracken fern is turning yellow, it is difficult to follow a herd as it courses along the flanks of some hill, even with a field-glass, on account of the resemblance between the general colour of their bodies and that of the ground on which our vision projects them. It would appear, therefore, that the colours of the bodies of these much-hunted animals may be more or less protective.

Nowhere in English literature will the reader find such capital descriptions of the habits and economy of the Red Deer as in the late Mr. St. John's "Wild Sports and Natural History of the Highlands." We are there introduced, in language worthy of Gilbert White, to the stag and his hinds, as they appear at home amid the rock-fastnesses of the Scottish hills and mountains.

One of the most interesting details concerning the stag is the growth of its antlers. As most people are aware (and as has already been referred to), these antlers are shed every year. Up to the stage of complete maturity each year has to produce a new growth, which is usually distinguished from its predecessors by additional *tines* or branches. When we remember the weight of these antlers, and that the solid matter composing them has to be abstracted from the blood and built up into these defensive forms within a few months, we see what an active physiological process must then be going on. Speaking of the antlers of the Red Deer, Prof. Bell says :— " By the number of these antlers, and other marks in the development of the horns, the age of the animal may be nearly ascertained. The growth of the horns is an astonishing instance of the rapidity of the production of bone under particular circumstances, and is certainly unparalleled in its extent in so short a period of time. A full-grown stag's horn probably weighs 24 lb. ; and the whole of this immense mass of true bone is produced in about ten weeks. During its growth the branches of the external carotid arteries, which perform the office of secreting this new bone, are considerably enlarged, for the purpose of conveying so large a supply of blood as is necessary for this rapid formation. These vessels extend over the whole surface of the horn as it grows, and the horn itself is at first soft and extremely vascular, so that a slight injury, and even merely pricking it, produces a flow of blood from the wound. It is also protected at this time with a soft, short hairy, or downy coat, which is termed *velvet ;* and hence the horns are said to be "in the velvet" during their growth. When completed, the substance of the horn becomes dense, the arteries become obliterated, the *velvet* dies and falls off in shreds—a process which is hastened by the animal rubbing his horns against the branches of a tree. The horns remain solid and hard, constituting the most effectual weapons of defence; and they are often used during the pairing season in violent and sometimes fatal combats between the males. After this season is over, absorption takes place at the point where the horn joins the boss or frontal process, and it at length falls off, to be renewed again in due time."

The Roebuck (*Cervus capreolus*), another and much smaller species of deer, is now almost restricted in its wild state to Scotland. It is naturally better adapted to a mountain life than the Red Deer, and indeed we may regard it as being "at home" only when found in such a habitat. Its bodily frame is lightly built, and it can leap from crag to crag with almost the agility of the chamois. The Roebuck is said to choose its female companion for life, and the affection which the couple manifest for each other and for their young is very remarkable. Both male and female, but especially the former, are quite as bold as the deer when defending their young, or when at bay; and the peculiar pointed tips of their horns are terrible stabbing weapons, often inflicting death on dogs, or even on the incautious hunter.

Roebuck, however, have latterly been shot down even in the Highlands, on account of the havoc they commit among the young plantations. Their flesh is rank, and of a goat-like flavour, and is conse-

quently little in demand even among the poor. Except
for its graceful shape, therefore, there is little reason
for its "preservation"; and the areas of Great Britain
where there is a possibility of its continuing wild in
spite of man's desires to the contrary, are every year
getting more constricted.   Mr. St. John, who was both

THE ROEBUCK (*Cervus capreolus*).

an ardent sportsman and a true observant naturalist,
writes of this creature as follows :—" Much as I like
to see these animals (and certainly the Roebuck is the
most perfectly formed of all deer), I must confess
that they commit great havoc in plantations of hard
wood.   As fast as the young oak-trees put out new

shoots, the Roe nibble them off, keeping the trees from growing above 3 or 4 feet in height by constantly biting off the leading shoot. Besides this they peel the young larch with both their teeth and horns, stripping them of their bark in the neatest manner imaginable. One can scarcely wonder at the anathemas uttered against them by proprietors of plantations. Always graceful, a Roebuck is peculiarly so when stripping some young tree of its leaves, nibbling them off one by one in the most delicate and dainty manner. I have watched a Roe strip the leaves off a long bramble shoot, beginning at one end and nibbling off every leaf. My rifle was aimed at his heart and my finger was on the trigger, but I made some excuse or other to myself for not killing him, and left him undisturbed—his beauty saved him. The leaves and flowers of the wild rose-bush are another favourite food of the Roe. Just before they produce their calves, the does wander about a good deal, and seem to avoid the society of the buck, though they remain together during the whole autumn and winter. The young Roe is soon able to escape from most of its enemies. For a day or two it is quite helpless, and frequently falls a prey to the fox, who at that time of the year is more ravenous than at any other, as it then has to find food to satisfy the carnivorous appetites of its own cubs."

Two of the rarest of our native mountain mammals are the Alpine Hare and the Wild Cat. They stand towards each other in as contrasted characters as the wolf and the lamb. One is the very type of trembling harmlessness, and the other is equally an illustration of sanguinary ferocity. The Alpine Hare (*Lepus variabilis*) derives its Latin specific name from its habit of changing the colour of its fur—a character which it has in common with its Arctic relatives. The manner in which it changes is as follows :—About the middle of September the feet, which before were of a grey colour, begin to turn white. Before the end of the month all the feet are of this hue. The white gradually gains on the grey, ascending the legs and thighs, whilst white hairs become more abundant on the belly. Up to the middle of October, however, the back still continues of a grey colour, although the ears and eye-brows are nearly white. From this period the colour-change proceeds very rapidly, and by the middle of November the whole of the fur, with the exception of the tips of the ears, is of a shining white. The singular fact about this annual change is that no *moult* takes place—not a hair falls from the animal ! Hence there can be no doubt that it is the hair which actually changes its colour. This whiteness lasts until the end of March, when it begins to turn to its grey colour again. By the middle of May, the animal has assumed that sober grey fur which is as protective for it in the summer as it is evident the dazzling white must be in the winter.

The Alpine Hare is most abundant on the mountains of Northern Scotland, although Professor Bell says it is occasionally found as far south as the hills of the Lake district of Cumberland. In size it is inter-mediate between the bulk of the common Hare and that of the Rabbit. It does not construct a burrow, like the latter animal, but hides in the clefts of the rock, or under stones. In the summer it will only be found at high elevations, but the rigorous cold of the winter drives it lower down, to feed on the lichens or seeds of pine.

The Wild Cat, or " Cat-a-mountain" (*Felis catus*) has been greatly hunted down and locally extirpated during the present century, on account of the immense havoc it commits among game-preserves and in poultry-yards. Few animals are more bloodthirsty or ferocious, and its great strength and agility render it sometimes even a dangerous foe to man. It inhabits only the most lonely and inaccessible ranges of crag and mountain, and is very seldom seen during the daytime. The poor Alpine Hare finds in it its chief and most implacable enemy. Mr. St. John, who had ample opportunities of observing the Wild Cat in its mountain home, writes of it in his note-book as follows :—"It prowls far and wide, walking with the same deliberate step, making the same regular and even track, and hunting its game in the same tiger-like manner ; and yet the difference between it and the domestic Cat is perfectly clear, and visible to the commonest observer. The Wild Cat has a shorter and more bushy tail, stands higher on her legs in proportion to her size, and has a rounder and coarser look about the head. The strength and ferocity of the Wild Cat when hemmed in or hard pressed are perfectly astonishing. The body when skinned presents quite a mass of sinew and cartilage. I have occasionally, though rarely, fallen in with these animals in the forests and mountains of this country ; once, when grouse-shooting, I came suddenly, in a rough and rocky part of the ground, upon a family of two old ones and three half-grown young ones. In the hanging birch-woods that border some of the Highland streams and lochs, the Wild Cat is still not uncommon, and I have heard their wild and unearthly cry echo far in the quiet night as they answer and

THE WILD CAT.

call to each other. I do not know a more harsh and unpleasant cry than that of the Wild Cat, or one more likely to be the origin of superstitious fears in the mind of an ignorant Highlander."

The Wild Cat formerly had a much more extensive distribution than it enjoys now; and there can be little doubt the introduction of firearms, in place of the old bows and arrows, has had a great deal to do with its local extirpation. But it is still met with, although rarely, in the elevated woods of northern England, as well as among the wooded mountains of North Wales, and in some parts of Ireland. The hilly copses of the Highlands, however, are the places where it is most commonly met with.

Prof. Bell speaks of this animal as follows :—"The strength and fierceness of this species are such as to render it an adventure of no trifling annoyance, and even of some danger, to come into close quarters with it, especially when it is exasperated by a wound. It is no pleasant affair to encounter an enraged male cat even of the domestic race; the strength and sharpness of his claws, and the length and power of his canine teeth, combined with a fierceness and rage which render such weapons doubly formidable, con- stitute him an opponent of no ordinary importance ; but the Wild Cat is still more to be dreaded from the greater size, power, and ferocity by which it is characterized. Hence Pennant designates it as the ' British tiger.' The female is considerably smaller than the male. She forms her nest either in hollow trees, or more commonly and more safely in the clefts of rocks ; and she has even been known, as Sir William Jardine says, to usurp the nest of some large bird as her own. She usually brings four or five young. The Wild Cat is found throughout the whole of those countries of Europe in which exten- sive forests exist, especially in Germany, and in all the wooded climates of Russia, Hungary, and of the North of Asia. These are of large size, and their fur is longer and held in much higher estimation than of those inhabiting warmer latitudes."

Several other bloodthirsty animals, now almost confined to hilly districts, belong to the Weasel family. The rarest of them is that called the Pine Marten (*Martes abietum*). Another goes by the name of the Beech Marten : both names are taken from the kinds of

THE PINE MARTEN.

forests they affect. The former has a yellow mark on the breast, and the latter a white one ; but Mr. St. John is of opinion that these are not two species, as he states he has seen individuals of every shade of intermediate difference in colour ; and he thinks the white-marked ones are the older animals. The female Pine Marten makes a nest of moss and leaves, but she generally usurps that of a squirrel if possible. This animal is not uncommon in the pine woods of Scotland, but it is because these trees grow well in the localities the Marten affects, rather than that it has a particular affection for this kind of tree. It is a great hunter of the White Hare in winter. The Highland shepherds accuse both it and the common Marten (*Martes foina*) of destroying sheep by seizing them at the nose, and gradually gnawing it away, so that the poor sheep die a lingering death of pain and hunger. All the martens live in trees, and herein differ from the usual habits of the other members of the Weasel family. The common Martens prey chiefly on small birds and squirrels, creeping from branch to branch in pursuit of them, although they frequently hunt on the ground in pursuit of hares, rabbits, rats, mice, &c. The common Marten is more particularly fond of mountain districts or rocky places; whence the origin of the other popular name it bears of "Stone" Marten.

The Ermine Weasel (*Mustela erminea*) is another of our native British mammals which changes its fur to white in the winter—a sure sign of its arctic origin and relationship. The whole body assumes a clear white, except the tip of the tail, which remains permanently black. Hence the robes of *Ermine*, formerly used only on state occasions, or as Royal furs, are remarkable for the jet-black spots on the pure white ground. This animal has been much sought after in

other countries, for the sake of its fur, the best, thickest, and whitest quality coming from more northern
regions.   The Ermine is not uncommon in hilly districts, particularly of Shropshire and North Wales.
It is still more abundant in the alpine districts of
Scotland ; and is met with on the hills and mountains
of the North of England.   Everywhere in these elevated regions its winter change of fur to white is complete ; but, singularly enough, when the Ermine occurs in the southern counties (which is a very rare in-
cident) such change is very partial indeed, and often
there is no difference whatever between the winter
and the summer fur.   The common name by which
this animal is known is the " Stoat," a corruption of
" Stout," in allusion to its being the largest of our
weasels.

The Pole-cat (*Mustela putorius*) is an allied species
to the Ermine, but too easily recognised by the much
more powerful and disagreeable odour it exhales,
especially when angry or when pursued by an enemy.
This creature is still far from uncommon in the north
of England, where it is commonly known by the
name of " Foumart."   It is more dreaded and hunted
down by the gamekeeper than any other of its race,
on account of its destructive and bloodthirsty habits.
When at bay not even the Wild Cat is more fero-
ciously obstinate and desperate.   In Smiles' " Life of
a Scotch Naturalist," we have a capitally told tale of
an adventure which Thomas Edward had with a
Pole-cat.   " One night, as he was lying upon a stone,
dozing or sleeping, he was awakened by something

pat-patting against his legs. He thought it must be a rabbit or a rat, as he knew they were about the place. He only moved his legs a little, so as to drive the creature away. But the animal would not go. Then he raised himself up, and away it went; but the night was so dark that he did not see what the animal was. Down he went again to try and get asleep; but before a few minutes had elapsed he felt the same pat-patting; on this occasion it was higher up his body. He now swept his hand across his breast, and thrust the intruder off. The animal shrieked as it fell to the ground- Edward knew the shriek at once. It was a Pole-cat.

" He shifted his position a little, so as to be opposite the doorway, where he could see his antagonist betwixt him and the sky. He also turned upon his side, in order to have more freedom to act. He had in one of his breast-pockets a water-hen which he had shot that evening; and he had no doubt that this was the bait which had attracted the Pole-cat. He buttoned up his coat to the chin, so as to prevent the bird being carried away by force." Edward tells the rest of the story himself. " Well, just as I hoped and expected, in about twenty minutes I observed the fellow entering the vault, looking straight in my direction. He was very cautious at first. He halted, and looked behind him. He turned a little and looked out. I could easily have shot him now, but that would have spoiled the sport; besides, I never wasted my powder and shot upon anything that I could take with my hands. Having stood for a few seconds, he slowly advanced, keeping his nose on the ground. On he came. He put his forefeet on my legs, and stared me full in the face for about a minute. I wondered what he would do next— whether he would come nearer or go away. When satisfied with his look at my face, he dropped his feet and ran out of the vault. I was a good deal disappointed; and I feared that my look had frightened him. By no means. I was soon reassured by hearing the well-known and ominous *squeak-squeak* of his tribe. It occurred to me that I was about to be assaulted by a legion of Pole-cats, and that it might be best to beat a retreat.

"I was just in the act of rising, when I saw my adversary once more make his appearance at the entrance. He seemed to be alone. I slipped quietly down again to my former position, and waited his attack. After a rather slow and protracted march, in the course of which he several times turned his head towards the door—a manœuvre which I did not at all like—he at last approached me. He at once leaped upon me, and looked back towards the entrance. I lifted my head, and he looked full in my face. Then he leaped down, and ran to the entrance once more, and gave a *squeak*. No answer. He returned. and leaped upon me again. He was now in a better position than before, but not sufficiently far up for my purpose. Down went his nose, and up, up he crawled over my body towards the bird in my breast-pocket. His head was low down, so that I couldn't seize him.

" I lay as still as death ; but being forced to breathe, the movement of my chest made the brute raise his head, and at that moment I gripped him by the throat. I sprang instantly to my feet, and held on. But I actually thought he would have torn my hands to pieces with his claws. I endeavoured to get him turned round, so as to get my hand to the back of his neck. Even then I had enough to do to hold him fast. How he screamed and yelled ! What an un-earthly noise in the dead of the night ! The vault rung with his howlings ! And then what an awful stench he emitted during his struggles ! The very jackdaws in the upper stories of the castle began to *caw*. Still I kept my hold. But I could not prevent his yelling at the top of his voice. Although I gripped and squeezed with all my might and main, I could not choke him.

" Then I bethought me of another way of dealing with the brute. I had in my pocket about an ounce of chloroform, which I used for capturing insects. I took the bottle out, undid the cork, and thrust the ounce of chloroform down the Foumart's throat. It acted as a sleeping draught. He gradually lessened his struggles. Then I laid him down upon a stone, and, pressing the iron heel of my boot upon his neck, I dislocated his spine, and he struggled no more. I was quite exhausted when the struggle was over. The fight must have lasted nearly two hours. It was the most terrible encounter I ever had with an animal of his class. My hands were very much bitten and scratched ; and they long continued inflamed and sore."

The Otter may be found at considerable elevations,
where the mountain-streams are constant enough to
harbour the trout and salmon on which it feeds. It
is frequently hunted in North Wales and the Lake district, as well as in Scotland. The Badger is still more
a hill-side animal, making its holes among the *débris* or talus of the mountains, and tracking the
neighbouring woods for roots, grass, or seeds. The
Rabbit is found on the sides of all the hills in
Wales and Cumberland, as well as those of Derbyshire, Lancashire, Yorkshire, Scotland, and Ireland.
In the latter country there is a species called the Irish
Hare (*Lepus Hibernicus*), with ears somewhat larger
than those of our English hare, although the head is
shorter. The fur also is different in its character and
quality.

The Fox may be regarded as another mountain
mammal, notwithstanding its abundance in the plains
and valleys even of our southern counties. Still, the
kind abundant on the mountains of the North of
England and Scotland is no doubt a well-marked
variety, adapted to mountain climbing, crawling, and
even leaping. It finds there abundance of animal
food, in the hares, rabbits, grouse, black-cock,
partridges, &c.

The most characteristic of these mammalia, however, are fast dying out ; and another generation or
two will witness the extinction of the Wild Cat and
the Pine Marten in .Scotland, just as the last eight
hundred years have comprehended the space of time
during which the Brown Bear and the Wolf have been

exterminated in the same country. Indeed, the slaughter of the last Highland Wolf occurred as recently as the year 1680. In Ireland, wolves were not entirely destroyed until about thirty years later. The dates are on historical record, as well as the names of the huntsmen who slew the last of those ancient inhabitants of our wooded mountains and open plains.

CHAPTER X.

OUR MOUNTAIN BIRDS.

The mountains a shelter for birds—The Erne—Golden Eagle—
Its Habits — Osprey — Buzzard — Kite — Peregrine Falcon
—Marsh Harrier—Hen Harrier—Short-eared Owl—Long-
eared Owl — Eagle — Owl — Lapland Owl — Ravens —
Ptarmigan — Black Grouse — Red Grouse — Capercaillie—
Golden Plover—Lapwing or Green Plover—Grey Plover—
Curlew—Common Snipe—Water-rail—Moor-hen.

THE rock-fastnesses of many English, Scotch, and
Irish mountains have formed capital retreats and
protective shelters for such birds as have strong
powers of flight, and therefore natural means of
reaching them. Elsewhere, in the valleys and plains
of the lowlands, where cultivation has utilized every
square mile of ground, raptorial birds have been all
but exterminated. A sparrow-hawk or kestrel, an
occasional merlin or owl, are all that remain of a
once splendid fauna of birds of prey. The preservation of game to such an extent has no doubt helped
to render such birds very rare ; and the reader will
bring vividly before his mind's eye the appearance
presented by the gable-ends of the buildings nearest
to gamekeepers' houses, where all the "vermin,"
birds as well as mammals, believed to be injurious to
game-preserving, were gibbeted and nailed up.

The preservation of nearly all our northern moors for grouse-shooting has carried the war of annihilation even among our hills and mountains ; and therefore birds of prey, which were once perhaps too abundant, are fast becoming as rare in these haunts as they have long been with us in the plains. Still, the solitude of the hills is the only place where the naturalist can hope to observe these wild birds in a state of nature. They have learned to be very wary of mankind, and the manner with which birds appear to understand the different ranges of improved guns and rifles, so as to just keep outside them, is a great proof of that wisdom which gains by experience. .

It is possible that among the mountains of northern Scotland and Ireland which are adjacent to the sea, particularly about Achil and near Westport in the latter country, we should see the Erne (*Haliaëtus albicilla*) grandly and slowly sailing above us, on the look-out for some stream or tarn where trout or salmon abound ; for it is a true fishing eagle, and never goes far from the sea. The Golden Eagle (*Aquila chrysaëtos*)—so called from the burnished-gold appearance of the feathers on its head and neck—is, however, our British type of eagle. It is certainly a more noble and majestic-looking bird than · any other of our native raptores. In the north-west of Ireland, as on Achil Island and in the mountains near the sea on that grand, wild Irish coast, the Golden Eagle perhaps may be found more numerously than elsewhere in Great Britain. It is also not uncommon in the mountains of Donegal. Still, it is occasionally seen in Derbyshire, and up to a recent date it bred among the mountains of Westmoreland and Cumberland, a crag near Keswick being the last place where we heard of an eagle's nest being known. It was kept a great secret, or a host of men and boys would have been in search of it. Golden eagles are still not unfrequent among the mountains of Perthshire and Sutherland, and the neighbourhood of Glencoe is noted for the numbers which abound. Ben Lomond, Ben Nevis, and other alpine heights, are also frequented by these noble-looking birds. Like the mountain-plants amidst which it loves to make its nest, the Eagle is an alpine bird. It is a mistake to suppose that the Eagle only feeds on what it has itself struck down. Mr. St. John gives several instances of its having been struck down by a stick when feeding and glutting itself on the carcases of dead or drowned sheep, and when it was so full that it was actually unable to rise. We quite agree with this author in the following remarks :—"It would be a great pity that this noble bird should become extinct in our Highland districts, who, notwithstanding his carnivorous propensities, should be rather preserved than exterminated. How picturesque he looks, and how perfectly he represents the *genius loci*, as, perched on some rocky point or withered tree, he sits unconcerned in wind and storm, motionless and statue-like, with his keen stern eye, however, intently following every movement of the shepherd or sportsman, who, deceived by his apparent disregard, attempts to creep within rifle-shot."

In the wild and romantic solitudes of Sutherland the Osprey or Fishing Eagle (*Pandion haliaëtus*) still breeds, and may be watched by the careful tourist who is provided with that capital adjunct, a good field-glass. The rocky islets in the mountain lakes are the favourite abodes and retreats of this now rare British bird. The Buzzard (*Buteo vulgaris*) can hardly be regarded as a specially mountain bird, for it is commonly distributed all over England. But the wildness of hilly and upland districts and its comparative freedom there from the gamekeeper, enable it to pair and nest in greater safety. The Kite (*Milvus regalis*) is rarely met with, and only in the hilly districts of North Wales and Scotland, whereas formerly it was distributed over nearly the whole area of Great Britain. It owes its general extirpation chiefly to gamekeepers, as it is a terribly destructive bird to grouse and other game. It appears to be an easily-trapped bird, on account of its greedy habits. The Peregrine Falcon (*Falco peregrinus*), once so much sought after and highly esteemed for falconry purposes, is also limited at the present time chiefly to the rocky districts of Scotland and Ireland. It has also been known to breed in the hilly regions of the West Riding of Yorkshire, and sometimes it is met with even in the south of England. It prefers, however, dizzy crags near the sea, for the sea-gulls constitute an important element in its dietary. Among the mountains both grouse and black-cock fall victims to its rapid flight. Our wild and marshy moorlands are sometimes visited by the Marsh Harrier (*Circus rufus*), whose slow, unsteady flight is soon identified by the ornithologist. Here also the allied species (more abundant in the meadows below),

called the Hen Harrier (*Circus cyaneus*), sometimes
puts in an appearance, to assist its cousin in hunting
for rabbits, moor-hen, or other water-fowl, mice, &c.

. Some of the owls are peculiarly mountain birds,
loving to hide in the daytime in the convenient

THE KITE (*Milvus regalis*).

clefts from the glare of the sunlight, and to issue
forth at night, when their hootings echo from the crags,
and deceive us with the idea that such solitudes must
be crowded with these remarkable birds of prey.
Their soft, delicate, noiseless flight, so different to

the whirr of a partridge, is produced by the specially constructed downy feathers of the wings. The Short-eared Owl (*Strix brachyotus*) breeds in the northern counties and in Scotland. The Long-eared Owl (*Strix otus*) must be looked for in the fir-woods of hilly districts of the North. This is a fine, bold bird, whose bright yellow eyes give to it a very handsome aspect. The markings and colour of the bird simulate those of branches on which it sits, so that only a close observation can detect it. The Eagle-Owl (*Strix bubo*) is an exceedingly rare visitant to the British isles, coming to us from the north of Europe. Although it has been shot (as most rare birds are in England, when they make their appearance) in various parts of this country, still the only places where we may expect to catch a glimpse of it are the Highlands. The lovely and stately Snowy or Lapland Owl (*Strix nyctea*)—a specially arctic bird, as its change of plumage to white in the winter plainly indicates—is only to be rarely met with in the rocky districts of the Orkney and Shetland islands, and occasionally along the eastern coasts of Scotland, whose proximity to Norway and Sweden, and the great prevalence of easterly winds, cause them frequently to be visited by rare birds.

Thomas Edward, the Scotch naturalist, thus writes concerning this bird:—"One of the most magnificent of the Owl tribe. What a splendid and showy bird! I think the term 'glistening' or 'spangled' might, with all truth and justice, be applied to this shining species. What a noble-looking bird! What beautiful eyes! the pupil dark, and the iris like two rings of

the finest burnished gold, set, as it were, in a casket
of polished silver.   I am proud of being able to give
this king of British owls a place in my list, and of
being able, perhaps for the first time, to say that at
least one pair have been known to breed in this
district.   A few miles west of Portsoy and not far
from Cullen, stands the bold and towering form of
Loggie Head.   In connection with this rocky pro-
montory, and about midway up its rugged height,
there is a narrow cave or chasm called ' Dickie Hare.'
In this cave a pair of these owls bred in 1845.   Un-
luckily, however, for them, a party of fishermen belonging to Cullen returning one morning from their
vocation discovered their retreat, by observing one of
the birds go in.   This was too good to lose sight of,
so up the dangerous and jagged precipice scrambled
one of the crew, and managed to reach the aperture
where the bird disappeared ; but instead of only one,
as he expected, he was not a little surprised to find
that he had four to deal with, two old and two young
ones well fledged ; and the apartment was so narrow
that only one person could enter at a time, so that
help was out of the question ; and his ambition
grasped the whole.   What was he to do, or what
could he do?   Turn?   Then the birds would have
flown.   No!   But just as I would have done had I
been in his place, he set upon them all ; and, after
a prolonged and pretty severe battle, in which he got
himself a good deal lacerated and his clothes torn by
the claws of the birds, he succeeded in capturing
them all alive, except one of the young ones, which
fell a sacrifice to the struggle."

The Ravens may be almost regarded as birds of prey, although they are more properly carnivorous birds, always on the look-out for wounded or dying animals. We may certainly regard them as peculiarly inhabitants of the mountains, although they are also abundant along the steep precipices of our northern seac-oasts. The reader may have seen them flying about Edinburgh Castle and Salisbury Crags. Wooded mountains are their favourite nesting-places, for there they are secure against attack. The Raven is another of our native birds which has become rare except in hilly districts. The introduction of firearms was the signal for its disappearance in the lowlands ; for the numerous references to it in our earlier writers show us how much more abundant it must formerly have been. Shakespere frequently alludes to the Raven.

But the birds which we naturally associate with our mountains and moors are those of the Grouse family. They are not to be found further south than the upland heaths of North Staffordshire, whilst their natural metropolis appears to be northern regions. This is certainly the case with the rarest of them (as far as Great Britain is concerned), the Ptarmigan (*Lagopus vulgaris*), whose white winter plumage plainly tells us it is an arctic bird, adapted to snowy wastes. It is abundant enough in Norway and Lapland, whence the Ptarmigan which now so cheaply stock our London markets are obtained. The name *Ptarmigan* is Gaelic, which would indicate that it must formerly have been much commoner in Scotland (to which part of Great Britain it is confined) than it now is ; for we uniformly find that when a natural object obtains a popular name, it is not uncommon. Indeed, this bird is believed to have once been an inhabitant of the uplands of North Wales, as well as of the mountains of the Lake district. At present it is confined to the Orkneys, the Hebrides, and a few places in the highest ranges

THE PTARMIGAN (*Lagopus vulgaris*).

of the Highlands. Mr. St. John gives us the following vivid description of this bird and its surroundings in his "Wild Sports and Natural History of the Highlands":—"Living above all vegetation, this bird finds its scanty food amongst the loose stones and rocks that cover the summits of Ben Nevis and some other mountains. It is difficult to ascertain what food the Ptarmigan can find in sufficient quantities on the barren heights where they are found. Being visited by the sportsman but rarely, these birds are seldom at all shy or wild ; but, if the day is fine, will come out from among the scattered stones, uttering their peculiar croaking cry, and, running in flocks near the intruder on their lonely domain, will offer, even to the worst shot, an easy chance of filling his bag. When the weather is windy and rainy, the Ptarmigan are frequently shy and wild ; and, when disturbed, instead of running about like tame chickens, they fly rapidly off to some distance, either round some shoulder of the mountain, or, by crossing some precipitous and rocky ravine, get quite out of reach. . . . The labour of reaching the ground they inhabit is great, and it often requires a firm foot and a steady head to keep the sportsman out of danger after he has got to the rocky and stony summit of the mountain. In deer-stalking I have sometimes come amongst large flocks of ptarmigan, which have run croaking close to me, apparently conscious that my pursuit of nobler game would prevent my firing at them. Once, on one of the highest mountains of Scotland, a cold, wet mist suddenly came on. We heard the Ptarmigan near us in all directions, but could see nothing at a greater distance than five or six yards. We were obliged to sit down and wait for the mist to clear away, as we found ourselves gradually getting entangled amongst loose rocks, which frequently, at the slightest touch, rolled away from under our feet, and we heard them dashing and bounding down the steep sides of the mountain, sometimes appearing, from the noise they made, to be dislodging and driving before them large quantities of *débris;* others seemed to bound in long leaps down the precipices, till we lost the sound far below us in the depths of the corries. Not knowing our way in the least, we agreed to come to a halt for a short time, in hopes of some alteration of the weather. Presently a change came over the appearance of the mist, which settled in large fleecy masses below us, leaving us, as it were, on an island in the midst of a snow-white sea, the blue sky and bright sun above us without a cloud. As a light air sprang up, the mist detached itself in loose masses, and by degrees drifted off the mountain-side, affording us again a full view of all around us. The magnificence of the scenery, looking down from some of these mountain-heights into the depths of the rugged and steep ravines below, is more splendid and awfully beautiful than pen or pencil can describe ; and the effect is often increased by the contrast between some peaceful and sparkling stream and green valley seen afar off, and the rugged and barren foreground of rock and ravine, where no living thing can find a resting-place, save the Eagle or Raven."

The wanderer on the northern moors and on the lovely mountain - sides has doubtless often been startled by the sudden rising of the magnificent Black-cock or Black Grouse (*Tetrao tetrix*), and by the loud cackling cries it emits when thus disturbed. It is, indeed, a splendid bird, fully deserving of all the high encomiums which sportsmen have lavished

upon it. Perhaps the observer may carefully steal
upon a flock of these birds in the early summer,
when the males are calling to each other and strutting
about like peacocks, trailing their wings like turkeys,
and going through such grotesque attitudes that he

THE BLACK GROUSE (*Tetrao tetrix*).

will scarcely forbear laughing outright. It is only the
male which ought to be accurately called "Black-
cock," as the plumage of the female is quite of a
different colour and markings, better adapted to
protect her from the keen outlook of birds of prey, as

she hatches her eggs. The tender shoots of the heather, as well as the ripe seeds of that plant, form their summer and winter food ; whilst in the autumn the berries of the Whortleberry, Bearberry, Cranberry, and many others, offer them a variety. Black Grouse are found on some high grounds in the south of England, as Dartmoor, Exmoor, and elsewhere ; but it is in the North, and particularly in Scotland, that they are most abundant.

A naturalist must live in the Highlands to thoroughly understand and sympathize with the animal life peculiar to those regions. No man ever felt the charm of such an occupation more than Mr. St. John. Speaking of the Black Grouse, he says, in his "Sketches" :—"During the spring, and also in the autumn, about the time the first hoar-frosts are felt, I have often watched the blackcocks in the early morning, when they collect on some rock or height, and strut and crow with their curious note, not unlike that of a wood-pigeon. On these occasions they often have most desperate battles. I have seen five or six blackcocks all fighting at once, and so intent and eager were they, that I approached within a few yards before they rose. Usually there seems to be a master-bird in these assemblages, who takes up his position on the most elevated spot, crowing and strutting round and round with spread-out tail like a turkey-cock, and his wings trailing on the ground. The hens remain quietly near him, whilst the smaller or younger male birds keep at a respectful distance, neither daring to crow, except in a subdued kind of voice, nor to approach the hens. If they attempt the latter, the master-bird dashes at the intruder, and often a short *mêlée* ensues, several others joining in it, but they soon return to their former respectful distances. I have also seen an old blackcock crowing on a birch-tree with a dozen hens below it, and the younger cocks looking on in fear and admiration."

The Red Grouse (*Lagopus Scoticus*) enjoys the distinction of being the only bird indigenous to Great Britain, and not met with anywhere else. The wildest of our heaths and moorlands are its favourite home ; and, as the reader is aware, it abounds on the Lancashire, Yorkshire, Westmoreland, and Cumberland moors. It is still more abundant in Scotland, and it is shot on the hills of North Wales, where it is said to attain its greatest size. The growing shoots of heather and the mountain berries form its chief food.

This bird is little aware of its large social and *political* importance, for there cannot be a doubt that many of the social arrangements of the upper classes in this country, as well as the time of concluding the summer session of Parliament, has reference to the 12th of August, when grouse-shooting commences. Mr. St. John thinks this date is *too early*, and says, " The Red Grouse generally make their nest in a high tuft of heather. The eggs are peculiarly beautiful and game-like, of a rich brown colour, spotted closely with black. Although in some early seasons the young birds are full grown by the 12th of August, in general five birds out of six which are killed on that day are only half come to their strength and beauty. The 20th of the month would be a much better day on which to commence their legal persecution. In October there is not a more beautiful bird in our island; and in January a cock grouse is one of the most superb fellows in the world, as he struts about fearlessly with his mate, his bright red comb erected above his eyes, and his rich dark brown plumage shining in the sun."

The Capercaillie, or Cock-of-the-Woods (*Tetrao urogallus*), has been reintroduced into Scotland, after having become extinct there. The name, which is Gaelic, shows us that it must have been tolerably abundant in former times. It also existed in a natural state in Ireland, from which country it disappeared rather more than a century ago. It is very abundant in the pine forests of Norway, whence it has been reintroduced into Great Britain. Perhaps it was the great size of this noble bird, and the well-known excellence of its flesh, which caused it to be hunted down until it became extinct. Its chief food appears to be the leaves and the tender young shoots of the Scotch fir, as well as those of the juniper. It also devours mountain berries, and, in winter, the buds of the birch-tree. In the pairing season the antics and attitudes of the Capercaillie are even more ludicrous than those of the Blackcock.

The boggy places on our hills and moorlands are the habitats where the Golden Plover (*Charadrias pluvialis*) most loves to frequent, for it is there that the worms, slugs, and insects which form their food are most likely to abound. This bird is not an uncommon object on our lonely moors, and the botanist often startles it from its bare nest on the ground

during his explorations. Like many others of its race,
the Golden Plover undergoes a change of plumage.
This is so remarkable in the Green Plover that it
goes by several different names expressive of the
change. The Golden Plover assumes its most attractive dress in the early summer, when the pairing season begins. Their shrill, whistling notes make the

THE LAPWING, OR GREEN PLOVER (*Vanellus cristatus*).

dreary moorlands appear all the more lovely. The
term "golden" may be in allusion to the triangular-
shaped spots of brilliant yellow which edge the feathers. The Lapwing (*Vanellus cristatus*), or Green
Plover (also known as the "Peewit," from its peculiar, breezy cry), usually swarms on our hilly heaths
and moors, where their insect food is plentiful. None

of our British birds is better known than this handsome species, whose lap-winged flight and shrilly calls
cannot fail to attract attention. The Grey Plover
(*Squatorala cinerea*) is a rare habitant of some of
the high grounds of Scotland. The Curlew (*Nume-
nius arquata*) is as common on the mountains near
the sea as the Peewit, and there can be little doubt
that its popular name is derived, like that of the latter
bird, from its peculiar cry. The boggy places on the
hills and moors furnish it with breeding-places, and it
is then that its cry seems to relieve the mountain solitudes. The Common Snipe (*Scolopax gallinago*) often
leaves the southern counties of England for the low
moorland heaths, where it loves to breed, in places
not far distant from the hill-side springs and rills
where aquatic insects and mollusca are abundant.
One of the common Gaelic names for it is the "Air-
Goat." Its bill is peculiarly formed for immediately
detecting its prey. The Water-rail (*Rallus aquaticus*),
as its name implies, frequents marshy and boggy
situations, where slugs and shell-fish are likely to be
met with. Its usual companion is the Moor-hen, and
the two birds are singularly adapted to the same conditions of life. It goes to the north only in the winter. The Moor-hen (*Gallinula chloropus*), as the
popular name imports, is very common among the
pools and tarns of our hills and moorlands, as indeed
it is everywhere if the necessary aquatic conditions
are present. Everybody is acquainted with the appearance of this common bird, and all have noticed
its quiet stealthy habits, well calculated to enable it
to glide away from its nest among the tall bulrushes

and sedges when an enemy threatens. They may
have seen them, early on some dewy summer morning,
jerkingly walking on the grass in search of their
breakfast of slugs and worms. In the water they nod
their heads first on one side and then on the other,
in rather a comical manner, whilst they are swimming
about.

WATER OR MOOR-HEN (*Gallinula chloropus*).

Shenstone refers to the habits of this bird, and its
frequent companion the Coot, in the following
lines :—

"To lurk the lake beside
Where Coots in rushy dingles hide,
And Moor-cocks shun the day."

Mr. St. John speaks of the Moor-hen as "often

leaving the water to feed with the poultry ; and walking about all day on the grass, with an air of the
greatest confidence and sensibility." He adds : " I
know nothing prettier than the young ones, as they
follow their parents in their active search for flies and
insects." Even in the Highlands the Moor-hens remain all the winter through. Formerly this bird was
called the " Mot-hen," because it was so abundant
in the neighbourhood of *moated* houses. No other
British bird was so soon reconciled to the appearance
and (to them) harmless character of railway-trains ;
and passengers may often see them feeding and
swimming in the dykes by the sides of railway embank-
ments without taking the slightest notice or manifest-
ing the smallest concern about the huge vehicle which
puffs and screams and roars its way past them !

CHAPTER XI.

OUR MOUNTAIN BIRDS (*continued*).

Nature's balance in the bird population—The Woodcock—Nut-cracker—Green Woodpecker—Cross-bill—Lapland Bunting—Snow Bunting—Siskin—Lesser Redpole—Mountain Linnet—Goldfinch—The Hawfinch—Reason for coloured fruits—The Brambling—Wheat-ear—Whin-chats and Stone-chats—Alpine Accentor—Redwing—Gold and Fire-crested Wrens—Alpine Swift—The Dipper, or Water-Ousel—The Dipper a modified Thrush.

THE wooded heights of our hills and mountains are frequently inhabited by birds which we should rarely meet with in the lowlands. Some of them affect Alpine conditions, inasmuch as they are in reality Arctic birds. Others seek these quiet shelters because they are so much less exposed to danger. And among the hills, Nature has a better chance of keeping her bird population in balance and order than she has in the plains, simply because man has not the same chance of interference. It is in the former places that we have the greatest quantity and variety of raptorial birds remaining, and these keep down the weakest of the tribes on which they prey. It is in the northern woods that the Woodcock (*Scolopax rusticola*) breeds; and it would appear as if it had

bred much more frequently since a larger area of
ground has been set with fir plantations. The Nut-
Cracker (*Nucifraga caryocatactes*) may possibly be still
seen in mountain woods, turning over the clinging
lichens in search of insects. The Green Woodpecker
is much more abundant and widespread. The common Cross-bill (*Loxia curvirostris*) particularly loves

THE CROSS-BILL (*Loxia curvirostris*).

the fir woods which fringe the hill-sides, or are
clustered as "cover" on the moors. Most of the
birds just mentioned climb like parrots. Mr. St. John
describes the habits of the Cross-bill as follows :—
"Whilst walking through the extensive fir and larch
woods I am often much amused by the proceedings
of those curious little birds the Cross-bills. They

pass incessantly from tree to tree with a jerking quick flight in search of their food, which consists of the seeds of the fir and larch. They extract these from the cones with the greatest skill and rapidity, holding the cone in one foot and cutting it up quickly and thoroughly with their powerful beak, which they use much after the manner of a pair of scissors. When the flock has stripped one tree of all its sound cones, they simultaneously take wing, uttering at the same time a sharp, harsh, chattering cry. Sometimes they fly off to a considerable height, and after wheeling about for a short time, suddenly alight again on some prolific-looking tree, over which they disperse immediately, hanging and swinging about the branches and twigs, cutting off the cones, a great many of which they fling to the ground, often with a kind of impatient jerk. These cones I conclude are without any ripe seed. They utter a constant chirping while in search of their food on the branches."

Some of the Buntings we meet with are characteristic mountain birds, and as two species at least are Arctic, it follows that we only get them in the north, unless when some exceptionally hard winter drives them to the eastern counties or the south of England, as was the case in the winter of 1878. The north-easterly winds then sometimes bring them over in flocks, and we get them all along the eastern and south-eastern coasts (as at Brighton), alighting in such a weakly condition that they are easily run down and taken by the hand. Sometimes the Lapland Bunting (*Plectrophanes Lapponica*) reaches us in this unceremonious and possibly unpremeditated manner.

The Snow Bunting (*Plectrophanes nivalis*) is of much more frequent occurrence, and Mr. St. John tells us that they visit the Highlands in large flocks, and remain there during the whole winter. The northern sea-shores are often covered with them. They usually arrive in October and November, and they are then

THE SNOW BUNTING (*Plectrophanes nivalis*).

of a much darker colour than they are a few weeks later on, when their plumage has gradually changed (as we have observed it does in all true Arctic birds and mammals) to that lighter and whiter hue which has given to the bird its name of the "Snow" Bunting. It is also called the "Tawny Bunting"; but this is only when the birds appear in their

brown summer plumage ; and we get them in every
transition stage from this to their perfect winter
dress. The tender buds of the opposite-leaved
saxifrage are said to furnish it with its favourite food,
and we have seen that this is one of our most
abundant upland plants, especially where there is
constant moisture. In Sweden the Snow Bunting
is only found in the summer time on the summits of
the high hills and mountains. It has been seen as
late as August on the tops of the highest mountains
in Scotland, where patches of the previous winter's
snow had remained unmelted. It is believed to
breed there and also on the Grampians.

The Siskin (*Carduelis spinus*) also comes to the
south-east and eastern counties during hard winters,
and we have frequently seen this graceful and tame
little bird swinging on thistles and other plants whose
seeds had not yet fallen from their capsules, and
engaged in diligently searching for them, passing in
short jerky flights from one plant to another, and
thus taking up much time in clearing a hundred
yards of ground. Both it and the Lesser Redpole
come to us in the late autumn from the north. It
is known to nest among the gorse bushes in the hills
of Scotland. The Lesser Redpole (*Linota linaria*)
is even a smaller bird than the Siskin. Only in
the north is it a summer resident, and we have
often found its nest, which is of the most exquisitely built character, formed of mosses and
lichens, and so perched in the forks of the branches
that we had to look long and closely before detecting
it. The underwood and copses which frequently

cover the lower flanks of the northern hills and mountains are the favourite nesting-places of this pretty and easily tamed bird. It is perhaps the favourite bird among the boys of North Lancashire and Yorkshire.

The Mountain Linnet (*Linota montium*) is a nearly allied species to the Redpole; from which bird,

MOUNTAIN LINNET (*Linota montium*).

however, it may be distinguished at a glance by the greater length of its tail. This, as Yarrell well says, gives to the Mountain Linnet a very elongated and gracefully slender appearance. We have frequently watched this bird in some of the loneliest spots in the elevated mountain regions of Scotland, solitarily passing from plant to plant of the common knap-weed, and picking out its seeds, causing the already top-heavy plant to sway to and fro by its additional weight. In the winter it leaves the mountains, and assembles in flocks in the lowlands; even migrating as far as the southern counties of England when the winter threatens to be a "hard" one; and thus the Mountain Linnet is one of those birds whose southern migrations are a sure token to the observant naturalist of the kind of winter which is likely to be experienced.

Far up the hill-sides, where the thistles and rag-worts are plentiful, we have met with the loveliest of our British birds, the Goldfinch (*Carduelis elegans*). The easy manner with which this bird has been taught to do "tricks" is a tribute to its superior intelligence. It has an intelligent look, very different from the "street Arab-like appearance," which the common house sparrow assumes even in the country. Unfortunately for the Goldfinch, its intelligence is only too likely to procure it a place in the narrow and dirty cage of some Whitechapel "rough" as a "draw-water"! Let us hope, however, that, as no man can be utterly bad who cares for birds and flowers, the captured goldfinch may be one of God's missionaries in many a hovel in the East End, where no other missionary is welcome.

The Hawfinch (*Coccothraustes vulgaris*) is not by any means so uncommon an English bird as some of the older ornithologists imagined. In some places in the eastern counties, in winter, we have known it to occur in such numbers as almost to call it abundant —notably at Yoxford, a village whose picturesque beauty has caused it to be not inaptly termed the

" Garden of Suffolk." As its name implies, this bird is particularly fond of the fruit of the hawthorn ; and in the winter of 1878 we watched five or six of these birds on a hawthorn-tree which had been allowed to grow out of the ordinary hedge to the height of about 15 feet, and which was laden with scarlet haws. The havoc they committed amongst this fruit (almost as much falling to the ground as was eaten), both proved to us that this bird was rightly called the Hawfinch, and also showed us what was the reason for the attractive colour of the haws themselves. One could see that the brilliant red of the latter made them easily visible to the eyes of hungry birds, and thus enabled the " stony"-environed seeds to be swallowed, carried to a distance by the birds, and then voided, but at a distance from the parent tree. So that this peculiar adaptation is quite as beneficial to the haw as if its seeds had been endowed with a feathery pappus like the seeds of the dandelion, by means of which the wind can transfer them to a distance, and thus secure them a wide-spread distribution.

But the Hawfinch is not such a strictly-confined " vegetarian" as to eat nothing but haws. It is fond of the *coloured* fruits of many other shrubs and plants, notably those of the hornbeam.

The Brambling, or Mountain Finch (*Fringilla mon-tifringilla*), is another companion of the solitary naturalist upon whom the " power of hills" is strong. In the southern counties it is known only as a winter visitor ; but we have seen it in the early spring and summer in the Derbyshire hills, looking out for such seeds of the previous year's knot-grass as had not yet been shed from their capsules.

The Wheat-ear (*Saxicola œnanthe*) is nearly allied to the two species of birds—the Whin-chat and the Stone-chat—which we naturally look for whenever we are travelling over ancient heathy or moorland ground. The Wheat-ears nest in the southern part of England,

THE BRAMBLING (*Fringilla montifringilla*).

generally making their nest of the common bracken fern on the downs or hill-sides, where they are only too likely to fall a prey to such shepherds as are induced by the temptation of "filthy lucre" to snare wheat-ears for the market which scientific gastronomers have called into existence.

The Whin-chat (*Saxicola rubetra*) is a well-known

and abundant bird on our commons and moorlands, and waste places generally. As its name implies, it is usually met with wherever the whin or gorse grows plentifully ; and we get into the habit of looking for this dainty bird perched on the top of almost every other bush. Like many others, it will feign immense concern in order to lure us away from its nest ; and its cry seems loudest and most excited

THE WHIN-CHAT (*Saxicola rubetra*).

when it is best succeeding in its innocent artifice. It darts in undulating flights from bush to bush, dropping for a second or two on the topmost twigs and swaying to and fro ; then rising into the air, and whistling its breezy note, it leads on a step further away from where its helpless little ones are nestling, ignorant of their mother's device. The Whin-chat is less hardy than the Stone-chat, and seldom passes

the winter in Britain; whereas the latter may be seen
on our heaths and moors nearly as plentifully then as
in the summer time, when the gorse and heather are
in bloom. The Stone-chat (*Saxicola rubicola*) is also
a much handsomer bird, and may be easily recognised
by the dusky neck and head, and the almost brilliant
white zone running from the chest to the shoulders

THE STONE-CHAT (*Saxicola rubicola*).

and the rich chestnut of the front part of the belly.
Its habits and even notes are very much like those of
its very near ally the Whin-chat; with the exception
that we shall find it at greater altitudes than the
latter.

The Alpine Accentor (*Accentor alpinus*) is only
likely to be found on the mountain-heights. The

Redwing (*Turdus iliacus*) is a lover of the rude
north, and nests amid the upland woods and forests;
although it often migrates to us in the south in the
winter time. Those lovely little birds the Gold-crested
Wren and the Fire-crested Wren are also much
commoner in the north than the south, where they
must be looked for in the lonely fir and larch plantations which cover the hill sides. The Alpine Swift
(*Cypselus alpinus*) builds its nest among the rocky
crags of the mountains, and wheels its tireless flight
around the precipices, stopping occasionally to cling
to their almost perpendicular faces for a moment or
two.

One of the commonest and most interesting of the
birds which the angler in mountain streams, or the
wandering botanist in search of rare plants, is likely
to see is the Dipper, or Water-Ousel (*Cinclus aquaticus*). At one moment it is beheld perched upon
some moss-covered stone which stands above the
water, and the next it has disappeared with a jerky
pitch of its tail. Although nearly allied to the
Thrushes, this bird assumes the habits of a Water-
hen. It can both dive and swim, and there is now
no doubt whatever that it is able to achieve the difficult feat of walking under water at the bottom of the
stream, in search of shell-fish, caddis-worms, and
perhaps fish-spawn. We have whiled away many a
summer hour among the hills in watching the habits
of this peculiar bird. It is a bad and ungraceful
walker, but if the observer will notice the feet he will
see the reason for this. The long, curved claws are
intended to grasp firmly the slippery stones in its

submerged rambles, and they are therefore specially
fitted for this habit. We can hardly expect that such
a thorough observer of mountain natural history as
Mr. St. John would do otherwise than specially note
this little bird. He says : "I do not know, among
our common birds, so amusing and interesting a little
fellow as the Water-Ousel. In the burn near this

THE WATER-OUSEL (*Cinclus aquaticus*).

place there are certain stones, each of which is always
occupied by one particular Water-Ousel; there he
sits all day, jerking his apology for a tail, and occasionally darting off for 100 yards or so, with a
quick, rapid, but straightforward flight; then down
he plumps into the water, remains under for perhaps
a minute or two ; and then flies back to his usual

station.   At other times the Water-Ousel deliberately
walks off his stone down into the water, and, despite
of Mr. Waterton's strong opinion of the impossibility
of the feat, he walks and runs about on the gravel
at the bottom of the water, scratching with his feet
among the small stones, and picking away at all the
small insects and animalculæ which he can dislodge.
On two or three occasions I have witnessed this act
of the Water-Ousel, and have most distinctly seen
the bird walking and feeding in this manner under
the pellucid waters of a Highland burn. .   .   .
The Water-Ousel has another very peculiar habit
which I have not heard mentioned.   In the coldest
days of winter I have seen him alight on a quiet
pool, and with outstretched wings recline for a few
minutes on the water, uttering a most sweet and
merry song ; then rising into the air, he wheels round
and round for a minute or two, repeating his song as
he flies back to some accustomed stone.   His notes
are so pleasing that he fully deserves a place in the
list of our song-birds.   In the early spring, too, he
courts his mate with the same harmony, singing as
loudly as he can.   Often have I stopped to listen to
him as he flew to and fro along the burn, apparently
full of business and importance ; then pitching on a
stone, he would look at me with such confidence
that, notwithstanding the bad name he has acquired
with the fishermen, I never could make up my mind
to shoot him.   He frequents the rocky burns far up
the mountains, building in the crevices of the rocks,
and rearing his young in peace and security, amidst
the most wild and magnificent scenery."

Hence it seems that, in spite of the extreme change of habits which the Water-Ousel has assumed, it has not yet lost the characteristic notes which make the Thrushes, of which it is a member, pre-eminent among song-birds.

CHAPTER XII.

THE INSECTS OF THE UPLANDS.

Relation between Alpine Plants and Alpine Insects—Relation
between Alpine Lepidoptera and Arctic Species—The
small Ringlet Butterfly—Marsh Ringlet—Rothlieb's Marsh
Ringlet — Scotch Brown Argus — Small Pearl-bordered
Fritillary — Northern Brown butterfly — Grayling — Small
Heath butterfly—Green Hair-streak—Night-flying Lepi-
doptera or Moths—The Lepidoptera of Rannoch Moor—
Rannoch Sprawler—Rannoch Looper—Cousin-german—Gray
Rustic — Yellow Underwing — Northern Swift — Netted
Mountain Moth—Gray Mountain Carpet Moth—Heath Rivulet
—Mountain Rustic—Pine Beauty—Black Mountain Moth—
Scotch Annulet—Welsh Wave — Smoky Wave — Welsh
Clearwing — Alpine Beetles, and other Insects — *Misodera
arctica*—Acidota—Bee-beetle—Conclusion.

THE presence on our mountains of so many species
of Alpine flowering plants, and the abundance of
others which love to grow on the hill-sides and the
breezy moors, would naturally suggest that we should
find peculiar insects, some of which would perhaps
not be found elsewhere. Now that we know the
purpose of colour and perfume in flowers is to attract
insects, and that of all flower-haunting insects, butterflies and moths are the most remarkable, it would
follow that specialized varieties of our common insects ought to be found on our hills and mountains,

able to bear the greater cold and dampness; or else
that peculiarly mountain species should be met with.

In a great measure this expectation, which, it should
be observed, is based entirely on the general harmony
and balance of compensatory arrangements which
exist in nature, is realized. We have both mountain
species, and mountain varieties of lowland insects.
Some of our native butterflies and more of our indigenous moths are only to be found in the northern counties, and this usually means that they are
distributed along the hill or mountain sides.

A few of these Alpine insects are nearly related to,
if not identical with, species found within the Polar

Upper Side.

Lower Side.

THE SMALL RINGLET (*Erebia epiphron* or *Cassiope*),

Circle during the recent Arctic expedition; thus
showing there is a similar relation between our

mountain insects and those of the extreme north, which we have already traced between the Alpine and Arctic flora.

One of the most remarkable of our mountain butterflies is the "Small Ringlet," usually known among collectors as *Cassiope* and *Epiphron*. It is a dark, dusky insect, which we have seen in considerable numbers high up along the flanks of Fairfield, near Ambleside. Its caterpillars feed on the mat-grass, and the sheep's fescue-grass. The Cumberland mountains are remarkable for their abundant possession of this otherwise rare insect. It has also been met with on that noble mountain Croagh Patrick, near

Upper Side.

Lower Side.

SMALL BORDERED FRITILLARY (*Argynnis Selene*).

Westport, and on that paradise in Scotland, for Alpine lepidoptera, Rannoch Moor. All the "Ringlet" butterflies are fond of hill-sides, and here in boggy places, at as great heights as 2,000 feet, we may find the "Marsh Ringlet"; or, in similar habitats, "Rothlieb's Marsh Ringlet." As the name implies, the "Scotch Brown Argus" is peculiar to that country, and will be found wherever the rock-rose grows, for its larvæ feed on that plant. On Rannoch Moor, Orrock-hill, and particularly in Perthshire, this remarkable butterfly may be collected. In moisty places along the northern and Scottish hill-sides, there also occurs that lovely little insect the "Small Pearl-bordered Fritillary." A well-marked variety of another wide-spread butterfly, the "Small Tortoise-shell," remarkable for being of larger size than its English relatives, may be seen flying everywhere to the very summit of Ben Lawers. The well-known and nearly cosmopolitan "Painted Lady" is another butterfly varietally adapted to our British mountains. Mr. Newman mentions having seen it flying near the stone cairn on the top of Snowdon, and it is found in the Highlands, from the sea-level to the top of Ben Lawers. The beautiful and much-prized "Northern Brown" is very common on Scottish mountains; but it is not found at such great altitudes as the "Marsh Ringlet," which latter butterfly the collector will perhaps most easily find on Rannoch Moor. The "Northern Brown" is particularly abundant in Inverness-shire. The hills of the Lake district, Westmoreland, North Lancashire, and of the West Riding are also visited by this insect. That magnificent butterfly the "Grayling" (*Satyrus Semele*) will perhaps be found by the naturalist on the stony moors at high elevations. It may be easily

observed, first settling on one flower and then on
another, seemingly as if more to bask in the sunshine

Upper Side of Female.

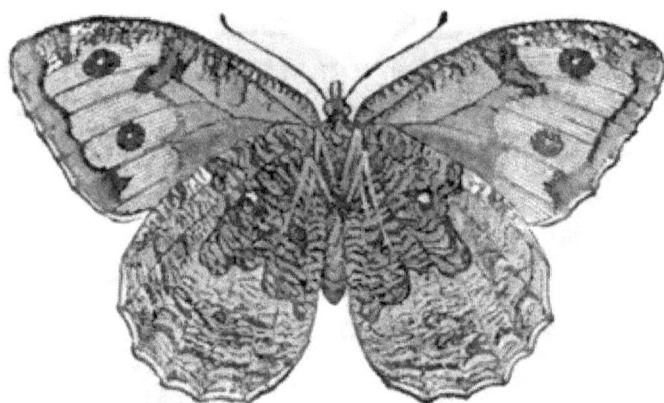

Under Side of Female.

THE GRAYLING (*Satyrus Semele*).

than to sip the nectar, although it may be doing
both.   It is not a timid butterfly, nor does it appear
to be fond of flight for its own sake, for it seldom
rises on the wing unless disturbed, and then only to
flutter away a short distance.   The "Small Heath"
is an abundant insect on most if not all our moors
and heathy hill-sides.   That lovely little butterfly

the " Green Hairstreak " (*Thecla rubi*) frequents the boggy places at the base and along the flanks of mountains, and is even taken at considerable altitudes, as for instance on Rannoch Moor, which is 1,500 feet above the sea-level.

Under Side.

Upper Side.

SMALL HEATH (*Cœnonympha Pamphilius*).

Our night-flying lepidoptera, popularly known as Moths, include a large number of species which, if not Alpine, are peculiar to hilly or mountainous districts. The locality last mentioned, Rannoch Moor, to reach which the tourist toils up the magnificent Pass of Glencoe, is a special area for several moths. As every one knows who has wandered over this moor, it is one of the wildest and most sparsely inhabited in the Highlands. Its surface is diversified by huge granite boulders, heathery patches, *sphagnum* bogs, steep-sided, sinuous, peat-discoloured streams and shallow lakes. Many a rare flower grows hereabout,

and doubtless the moths find them out for food-
plants.

Two species of British moths derive their popular

Upper Side.

Lower Side.

GREEN HAIRSTREAK (*Thecla rubi*).

names from this wild moorland. It should be
remarked that the nomenclature of our lepidoptera,

THE RANNOCH SPRAWLER MOTH (*Petasia nubeculosa*).

both scientific and popular, is in a very loose and
ill-defined condition. But we give their names as the
reader will find them in Newman's work, than which

we do not know a better nor more trustworthy.   The
woodcuts there given are quite sufficient to enable a
young entomologist to determine for himself the
species of any butterfly or moth he may have captured.   On the moor in question we find the "Rannoch Sprawler" (*Petasia nubeculosa*), a large and very
beautifully-marked moth.   Its caterpillars feed on the
leaves of the birch, the only tree, except a few Scotch
firs, which grows on the moor.   As yet this insect has
not been found anywhere else in the British islands
except on Rannoch.   On this and other Scotch
moorlands where the Bilberry grows, we should also
be likely to find one of the *Geometers* called the
"Rannoch Looper."   Another eccentrically named

THE RANNOCH LOOPER, or GEOMETER (*Fidonia pinetaria*).

moth, only met with on Rannoch, is the "Cousin-
german" (*Noctua sobrina*).   The "Gray-Rustic" is

THE COUSIN-GERMAN (*Noctua sobrina*).

also common there ; and this insect is also met with
on other moors and heaths, for its larvæ (which are

night-feeders) live on the tender shoots of the ling. The small dark "Yellow Underwing" (*Anarta cordigera*), a small, but lovely and elegantly-shaped moth, is also found only on Rannoch Moor, so far as Great Britain is concerned. That scarce *noctua* the "Broad-bordered White Underwing" is taken in the same locality; as is also the very rare "Slender-striped Rufous" moth.

Among the northern moths to be sought after by the collector on the Yorkshire and Scottish hills and moorlands is the "Saxon," another of the *noctuas*. The "Glaucous Shears" (*Hadena glauca*) is commoner in Scotland than elsewhere, although it may be found where the goat-willow grows, in Lancashire, Westmoreland, the Lake district, and elsewhere. The caterpillars of the "Northern Swift" (*Hepialus velleda*) feed on the underground branches of the bracken fern; and this moth is usually abundant in the North of England and Scotland, where the hillsides are covered with brake. It is a remarkable but not an attractive insect, which, as the popular name implies, has a rapid and jerky flight. The "Netted Mountain Moth" is only found on the hills

THE NETTED MOUNTAIN MOTH (*Fidonia carbonaria*).

of Yorkshire and in Scotland, generally in marshy places. It is a graceful insect, whose wings are agreeably and minutely freckled with black and white, in a kind of "pepper-and-salt mixture." The "Gray Mountain Carpet" moth lays its eggs on the stalks of the Whortleberry, and the young naturalist ought to look there for them on the moors of Durham, Lancashire, Yorkshire, the Lake district, Scotland, and Ireland. This is one of the loveliest of our mountain

THE GRAY MOUNTAIN CARPET MOTH (*Larentia cæsiata*).

moths, especially as regards the patterns or markings which give to this group the name of "carpet." The caterpillar is also noticeable for the delicate tints and colours employed in the ornamentation of its body. The "Yellow-winged Carpet" is another elegantly-marked moth, as regards the "patterns" on its fore wings. It is only found in the neighbourhood of the Lancashire hills, and those of the Lake district and Scotland, where the White Meadow Saxifrage grows. The "Heath Rivulet" is a tiny

THE HEATH RIVULET MOTH (*Emmelesia ericetata*).

insect found on hill-sides in Cumberland, Scotland, and on the Mourne mountains in Ireland.

Where the Sweet Gale (*Myrica gale*) shrub grows

in the spongy, boggy places of our moors, we may expect to find a few species of moths whose larvæ feed on the aromatic leaves of that tree. Among these are the grubs of the "Argent-and-Sable" (*Melanippe hastata*), a black-and-white insect. The caterpillars of another mountain species, the "Antler" (*Charæas graminis*), are very destructive to the grass on which they feed. An entomological writer speaks of it as follows :—"Some years ago, during the spring and early summer, the herbage of a large portion of the level part of the mountain Skiddaw, comprising at least fifty acres, was observed, even from the town of Keswick, to assume a dry and parched appearance; and so marked was the line, that the progress made by the caterpillars down the mountain could be distinctly noted. Nor was the change of colour of the herbage the only thing that attracted the attention of the good folks of Keswick; large flocks of rooks, attracted, no doubt, by the abundance of food which these caterpillars afforded them, were every morning seen wending their way to the spot; and after spending the day in preying upon the unfortunate caterpillars, on the approach of night, rising in one dense cloud, and dispersing to their respective homes. Though the number of caterpillars must in this manner have been greatly reduced, yet I was informed by a very intelligent friend residing at the foot of the mountain that in August the moths literally swarmed throughout the neighbourhood. So completely was vegetation destroyed that, on a visit to the spot more than a year afterwards, the extent of these ravages was distinctly visible, being very similar to the effect produced by the burning of the heather, which is so much practised on all our hills."

The "Mountain Rustic" (*Pachnobia carnica*) is an

THE MOUNTAIN RUSTIC MOTH (*Pachnobia carnica*).

exceedingly rare moth, which has only been taken in Scotland. The "Suspected" (*Orthosia suspecta*) is commoner, but still a very rare insect, met with only in the Lancashire and Yorkshire hills, and on those of the Lake district.

The "Pine Beauty" must be looked for in plantations of Scotch fir; where also the collector will be likely to find the eggs of the "Bordered White" moth (an exceedingly pretty insect, with large feathery antennæ). The "Bordered White" (*Fidonia piniaria*) is not an uncommon moth in Scotch-fir woods; and

THE BLACK MOUNTAIN MOTH (*Psodos trepidaria*).

the student will find its oblong eggs laid in rows on the needle-shaped leaves of this well-known tree.

The "Black Mountain Moth" is only found in the Highlands, where its smoky-black wings would soon help in its identification. The Scotch Annulet (*Dasydia obfuscata*) is another inhabitant of the

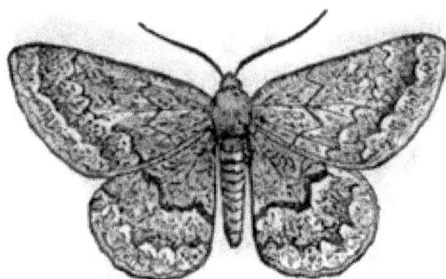

THE SCOTCH ANNULET MOTH (*Dasydia obfuscata*).

Scotch mountains, as well as those of Wicklow, in Ireland. Its caterpillars feed on the Dyer's Green Weed (*Genista tinctoria*), so that it should be sought for where this conspicuous plant is growing. The "Mirror Shoulder-knot" (*Epunda viminalis*) is also a Highland moth, although found elsewhere. Its larvæ may be found feeding on several species of willow. The inconspicuous but elegant little "Autumnal Moth" will only be met with in the north of England and Scotland.

THE WELSH WAVE (*Venusia Cambricaria*).

Another small but, among entomologists, well-known moth is the "Welsh Wave" (*Venusia Cam-*

bricaria). Its eggs are usually laid on the leaves of
the mountain-ash, or "rowan" tree, and there the
caterpillars feed and pass their chrysalid stage.
This insect may be found anywhere among the hills
of the northern English counties where the mountain-ash is abundant. The "Smoky Wave," an allied
species, is also northern in its distribution, being most
abundant, however, on heaths.

In the neighbourhood of Llangollen, North Wales,
entomological tourists always look out for the Welsh
Clear-wing (*Sesia scoliæformis*), one of that group of

THE WELSH CLEAR-WING (*Sesia scoliæformis*).

moths (whose popular name alludes to their partially
unscaled membranous wings) which so accurately
simulate the shape of stinging insects of quite another
order (the *Hymenoptera*), such as bees, wasps, hornets,
ichneumons, &c., that a young naturalist is easily
deceived by the remarkable resemblance. The caterpillars of the "Welsh Clear-wing" (which has only
been found at the above-mentioned locality) feed on
the wood of the birch-tree. We may, therefore,
claim it for a mountain insect. The almost *clear-*
winged "Narrow-bordered Bee Hawk-moth" (*Macro-*

glossa bombyliformis), as the scientific name implies, so much resembles a humble bee that it is only by taking a good look at its antennæ, as it darts past us, that we are satisfied of its lepidopterous character. Its caterpillars (which are rare) should be looked for on the Field Scabious in the northern counties of England.

Of course there are also gnats, two-winged or *dipterous* flies, as well as bees and ants, which frequent our hills and moors, especially where it is damp and boggy. These form the staple food of such upland birds as are insectivorous in their diet. A few species of beetles are also peculiar to our mountainous districts, and one or two of them are regarded as "prizes" by enthusiastic coleopterists. Perhaps the most important of these is the beetle called *Misodera arctica*, whose very specific name shows us its geographical affinities. It may be found on Snowdon, both on the Beddgelert and Pen-y-gwryd sides, but never far from the summits. The entire insect is only about $\frac{1}{3}$ inch in length ; and its colour is usually of a brown, with a brassy lustre, sometimes passing into greenish or bluish black. It occurs under stones on most of the high mountains of Scotland, as well as on the highest parts of the Lancashire, Yorkshire, and Derbyshire moorlands. A careful search under the stones near the summit of Snowdon and other Welsh and Scotch mountains will be repaid by the discovery of other rare beetles which would not be met with in the plains below. The moss-covered bark of the trees in northern fir-plantations also yields small beetles which cannot be found elsewhere ; such as *Acidota crenata*, &c. A very remarkable beetle, rarely to be seen in the south of England, although common in the north and in Scotland, is that known in the latter country by the name of the " Bee-beetle " (*Trichius fasciatus*). This beetle has a bee-like appearance, particularly about the thorax and hinder part of the body, which parts are covered with yellow hairs. The two yellow bands which run across the black elytra assist in the mimicry, even when the insect is at rest ; but when it is flying about the thistle flowers, as is its habit, it simulates the appearance of a humble bee to a very remarkable degree.

One of the rare finds on Rannoch Moor is *Hemerobius inconspicuus*, a neuropterous insect about whose life-history our scientific entomologists would be glad to know more, and which want might be supplied by the observant tourist.

We have now completed our task, which has been briefly to set forth the inorganic and organic history of our mountains, hills, and moors. It will have been seen how remarkably special physical conditions, such as these, are responded to by adaptations of the life which is found among them. It is this constant adaptation, always varying with the animals and plants as the conditions vary, which so deeply impresses us with the idea of an Untiring Wisdom, which " neither slumbers nor sleeps " !

Geological science shows us how the mountains gradually pass away or are brought low, and the valleys ultimately become exalted—the life of one period reaches its maximum and passes away, to be succeeded by another set of vital forms; but, throughout this ceaseless circulation of organic and inorganic forces, the same Intelligence keeps watch, and directs them all to that End unto which Creation itself, mighty though it be, is only a means!

THE END.

WYMAN AND SONS, PRINTERS, GREAT QUEEN STREET, LONDON.

s. d.

Beauty in Common Things. Illustrated by 12 Drawings from Nature, by Mrs. J. W. Whymper, and printed in Colours, with descriptions by the Author of "Life Underground," &c. 4to. ....*Cloth boards* 10 6

Botanical Rambles. By the late Rev. C. A. JOHNS, B.A., F.L.S. With illustrations and woodcuts. Royal 16mo..*Cloth boards* 2 0

Flowers of the Field. By the late Rev. C. A. JOHNS, B.A., F.L.S. With numerous woodcuts. Fcap. 8vo... *Cloth boards* 5 0

Wild Flowers. By ANNE PRATT, Author of "Our Native Songsters," &c. With 192 coloured plates, in two volumes. 16mo. ....*Cloth boards* 16 0

Forest Trees (The) of Great Britain. By the Rev. C. A. JOHNS, B.A., F.L.S. New Edition. With 150 woodcuts. Post 8vo. .... *Cloth boards* 5 0

Natural History of the Bible (The). By the Rev. CANON TRISTRAM, Author of "The Land of Israel," &c. With numerous illustrations. Crown 8vo. *Cloth boards* 7 6

Animal Creation (The). A popular Introduction to Zoology. By THOMAS RYMER JONES, F.R.S. With 488 woodcuts. Post 8vo. .... *Cloth boards* 7 6

Lessons from the Animal World. By CHARLES and SARAH TOMLINSON. With 162 woodcuts, in two volumes. Fcap. 8vo. ....*Cloth boards* 4 0

Birds' Nests and Eggs. With 22 coloured plates of Eggs. Square 16mo. ....*Cloth boards* 3 0

British Birds in their Haunts. By the late Rev. C. A. JOHNS, B.A., F.L.S. With 190 engravings by Wolf and Whymper. Post 8vo. ....*Cloth boards* 10 0

s. d.

British Animals. With 12 coloured plates. 16mo.
Ornamental covers   1   6

Birds of the Sea-shore. With 12 coloured plates.
16mo. .... *Cloth boards*   1   8

Evenings at the Microscope; or, Researches among
the Minuter Organs and Forms of Animal Life. By
PHILIP HENRY GOSSE, F.R.S. A new Edition revised
and annotated. With 112 woodcuts. Post 8vo.
Cloth boards   4   0

Familiar History of British Fishes. By FRANK
BUCKLAND, Inspector of Salmon Fisheries for England
and Wales. With a Frontispiece and 134 woodcuts.
Crown 8vo. .... *Cloth boards*   5   0

Natural History (Illustrated Sketches of); consisting
of Descriptions and Engravings of Animals. With
numerous woodcuts, in 2 vols. Fcap. 8vo.
Series I. and II. .... *Cloth boards, each Vol.*   2   6

Our Native Songsters. By ANNE PRATT, Author of
" Wild Flowers." With 72 coloured plates. 16mo.
Cloth boards   8   0

Selborne (The Natural History of). By the Rev.
GILBERT WHITE. With Frontispiece, Map, and 50
woodcuts. Post 8vo. ....*Cloth boards*   2   6

Ocean (The). By PHILIP HENRY GOSSE, F.R.S,.
Author of " Evenings at the Microscope." With 51
illustrations and woodcuts. Post 8vo. *Cloth boards*   4   6

Dew-drop and the Mist (The): an Account of the
Phenomena and Properties of Atmospheric Vapour in
various parts of the World. By CHARLES TOMLINSON,
F.C.S. With woodcuts and diagrams. Fcap. 8vo.
Cloth boards   2   6

s. d.

Frozen Stream (The): an Account of the Formation and Properties of Ice in various parts of the World. By CHARLES TOMLINSON. With woodcuts and diagrams. Fcap. 8vo. ....*Cloth boards* 1 6

Rain-Cloud and Snow-Storm: an Account of the Nature, Formation, Properties, Dangers, and Uses of Rain and Snow. By CHARLES TOMLINSON. With numerous woodcuts and diagrams. Fcap. 8vo. *Cloth boards* 2 6

Tempest (The): an Account of the Origin and Phenomena of Wind in various parts of the World. By CHARLES TOMLINSON. With numerous woodcuts and diagrams. Fcap. 8vo. ....*Cloth boards* 2 6

Thunder-Storm (The): an Account of the Properties of Lightning and of Atmospheric Electricity in various parts of the World. By CHARLES TOMLINSON. With numerous woodcuts and diagrams. Fcap. 8vo. *Cloth boards* 2 6

Winter in the Arctic Regions and Summer in the Antarctic Regions. By CHARLES TOMLINSON. With two maps, and several illustrations and woodcuts. Crown 8vo. ....*Cloth boards* 4 0

Depositories :

77, GT. QUEEN STREET, LINCOLN'S INN FIELDS, W.C. ;

4, ROYAL EXCHANGE, E.C. ; & 48, PICCADILLY, W.,

LONDON.

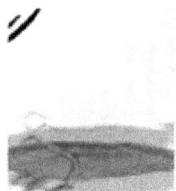